Vortex Publishing LLC.
4101 Tates Creek Centre Dr
Suite 150- PMB 286
Lexington, KY 40517

www.vortextheory.com

© Copyright 2019 Vortex Publishing LLC.

All rights reserved. No part of this book may be reproduced or transmitted in any form or by any means, electronic or mechanical, including photocopying, recording or by any information storage and retrievable system without the prior written permission by the Publisher. For permission requests, contact the publisher.

Printed in the United States of America

1 2 3 4 5 6 7 8 9 10

Library of Congress Control Number: 2019949630

ISBN   978-1-7332996-0-2
eISBN 978-1-7332996-1-9

Editor's note: All drawings in this book are original illustrations made by Dr. Moon. They are kept as they are to maintain the integrity of his work.

TABLE OF CONTENTS

Shocking Commentary by Prof., Dr. Victor V. Vasiliev II
Author's Note to the Reader IV
Introduction V

PART I

Chapter 1: A Strange Story in *Readers Digest* 1
Chapter 2: Where Is the Kingdom of God Located? 5
Chapter 3: The Search for the Kingdom of God Begins 9
Chapter 4: Higher Dimensional Space 11
Chapter 5: The Greatest Story That's Never Been Told 16
Chapter 6: The Five Pieces of the Universe 19
Chapter 7: Hard Times 25

PART II

Chapter 8: Einstein's Mistake 29
Chapter 9: The Curious Relationship Between Time and Motion 33
Chapter 10: THE MICHELSON MORLEY EXPERIMENT 37
Chapter 11: Bent and Flowing Space 42
Chapter 12: "Less Dense, and Flowing Space" 44
Chapter 13: Protons and Electrons Are Holes in Space! 46
Chapter 14: I Make the First Great Breakthrough 48
Chapter 15: I Discover How Atoms Are Really Constructed 50
Chapter 16: Energy Is Finally Explained Too 52
Chapter 17: The Secret of the Neutron 54
Chapter 18: The Four Forces of Nature Are Finally Explained 57
Chapter 19: The Shocking Truth About Force! 60
Chapter 20: At Last – The Alpha and the Omega 63
Chapter 21: Is This God? Have We Discovered God? 65

PART III

Chapter 22: The Search for a Proof Begins 68
Chapter 23: What Does It All Mean? 71

PART IV

Chapter 24: The Secret of Time Is Finally Explained 73
Chapter 25: Time Dilation Is Finally Explained 76
Chapter 26: The End of the Theory of Relativity 78

Epilogue 81
Appendix 84
The Thesis of the Mathematical Proof Is Presented 85

References 113

Subjects found in Book 2 119
Subjects found in Book 3 120

Shocking Commentary by Prof., Dr. Victor V. Vasiliev

Of all the millions of books that have ever been written throughout the history of mankind, only a handful have had the rare and distinct privilege of being able to change the way the entire human race forever thinks – this is one of them.

The reason for this extraordinary statement comes from the fact that this book introduces the shocking revelation that "time" – time itself – does not exist. This unprecedented discovery possesses implications of astounding proportion. Not only for all of science, but also, for all philosophy and religion. Imagine waking up and discovering every science text in the world is obsolete and every science course taught in every university in every country of the world is obsolete. NO! Except for the revolutionary works of Nicholas Copernicus – *De revolutionibus orbium coelestium* – there has never been anything like this book – ever!

When Mr. Moon originally discovered this shocking and frightening revelation many years ago, he was ignored by Western scientific community because he did not possess a PhD in Physics. Many American scientists would not even look at or read his mathematical proof! But all that has now changed. Through a remarkable series of events, Mr. Moon's discovery was introduced to the Russian scientific system: the system that possesses what can only be called: "the true scientific attitude".

The Russian scientific attitude is bold and revolutionary. Contrary to the attitude of the West, in Russia, an idea is examined first, and the individual's scientific credentials are examined second: exactly opposite to what they are in the America. Using this system, when Mr. Moon's ideas were examined, they were instantly recognized for the extraordinary discovery they truly are.

This theory of Russell Moon first came to my attention in 1992 while reading the abstracts of a conference he spoke at in Fort Collins, Colorado. Then Chairman of the Electro-Physics department at the *Lenin All-Russian Electrotechnical Institute* in Moscow, I was intrigued by its combination of simplicity and profound shocking implications. I contacted Mr. Moon and obtained a copy of the original paper.

I next revealed it to colleagues, and it eventually found its way into a number of other universities. After much excitement and discourse, including the publication of a pamphlet by Dr. Leonid Samuilovisch Slutskin of the "Publishing House ZNACK" of the MOSCOW POWER INSTITUTE TECHNICAL UNIVERSITY, articles on this remarkable discovery were eventually published in *Membrana,* a Russian scientific journal. These articles created much controversy.

These publications eventually drew the attention of Professor Konstantin Gridnev, one of Russian's greatest scientists. Professor Gridnev is a Theoretical Nuclear Physicist; author of books and numerous articles on nuclear physics [25 in the last 5 years alone]; Chairman of numerous world scientific conferences sponsored by Russian government; and Chairman of Nuclear Physics Department at the "Harvard of Russia" – St. Petersburg State University, in Petergof [The world-renowned university founded by Peter the Great in 1724.]

When Professor Gridnev read Mr. Moon's papers and examined the mathematics, he was so shocked by what he saw, he came to visit me in Moscow. Then in an unprecedented move, he came to America to spend two weeks with Mr. Moon to better understand this revolutionary theory and to see just what it is capable of explaining in the field of Nuclear Physics.

This trip was well worth it because he challenged Mr. Moon to use this theory to explain many of the great mysteries and conundrums of science that until this time, have had no explanation whatsoever. Although it is hard to believe, so far, over 80 of these great mysteries of science –

(listed in the back of this book) in Newtonian physics, Relativity, Quantum Mechanics, and Nuclear Physics – are explained: any one of which would have awarded its author a PhD in Physics.

Some of these explanations have now been published in over 30 international scientific conferences throughout the world. Many other papers are being held back from publication because of confidential research in the new phenomenon of quantum entanglement – another recently discovered scientific mystery this theory easily explains.

Because of his scientific papers, scientific articles, previous education and research, expansion of the original thesis to include the explanation for exactly how phenomenon of time itself is created; and a revolutionary scientific experiment performed at St Petersburg State University in 2005, [subsequently entitled The Photon Acceleration Effect]; Mr. Moon qualified for, and after a lengthy refereeing process, was eventually granted PhD in Nuclear Physics. The PhD was awarded in Moscow on October 19th, 2005, following resolution #7 of the Dissertation Council of St. Petersburg State University of June 30, 2005. The degree called THE ACADEMIC DEGREE OF CANDIDATE OF PHYSICAL-MATHEMATICAL SCIENCES, is equivalent to the PhD degree awarded by western universities and is given by the Russian Ministry of Education. [In Russia, higher academic degrees are awarded by government and each individual is assigned a number: KT#032771.]

This PhD now gives Professor Moon's discoveries listed in this book the backing of the Powerful Russian Ministry of Education. The list of Physicists that back this work is also impressive, especially those who are either the Chairmen or Leading Engineer of their departments in their individual universities. There has never been anything like this. Some of these distinguished scientists are: Prof., Dr. Lev Semenovich Ivlev professor of Physics and Mathematics, Chairman of the Aerosol Physics Dept. at *St. Petersburg State University*, in Petergof [author of 17 books, and 70 scientific papers]; Prof., Dr. Vladimir N. Puchinskiy of the *Moscow State Technical University* (MAMI); Dr. Alexander E. Filippov Leading Engineer of the V.I. *Lenin All-Russian Electro Technical Institute* (VIE); Dr. Alexander N. Vasil'iev Leading Engineer *of the Moscow Power Energy Institute (Technical – University)* (MEI); Prof., Nikolay Yur'evich Terekhin, Leading engineer Dept. of Atmospheric Physics, *St. Petersburg State University*, Petergof; Dr. Stanislaw I. Gusev, Chief of department, at the V.I *Lenin All-Russian Electro Technical Institute* (VIE). And of course myself - Prof., Dr. Victor Vasiliev, Chairman of the Electro-Physics department at the *Lenin All-Russian Electro Technical Institute* in Moscow; and Prof., Dr. Konstantin Gridnev Chairman of the Nuclear Physics Dept. at *St. Petersburg State University* in Petergof.

Prof., Dr. Victor V. Vasiliev; Jan. 2009

AUTHOR'S NOTE:

Although this book introduces perhaps the most important scientific-religious discovery ever made, it is written so everyone can understand it.

Divided into three parts,

PART I: explains why a construction worker went in search of the ultimate secrets of both science and religion.

PART II: EINSTEIN'S MISTAKE explains what was found, how it was found, and its astonishing implications. No scientific knowledge or mathematical skills are required.

PART III: the third part of this book tells about the search for a proof and what is creating the phenomenon of time.

THE APPENDIX originally contained a mathematical proof. This proof was my thesis for which I was awarded a PhD in Nuclear Physics by the Russian Ministry of Education in 2005. The thesis is now presented in Book 2 of this six part series.

(Note: if anyone who is not a physicist or a mathematician wants to read the proof, it should be mentioned that only a High School level of math is needed. Any High School Senior who has taken Trigonometry possesses the ability to understand this extraordinary proof. No PhD is necessary.)

The appendix also contains the discoveries contained in the next two books in this six part series.

In honor of two of Russia's greatest scientists: Dr., Prof. Konstantin Gridnev; and Dr., Prof. Victor V. Vasiliev, I have given the vortices the name "Konsiliev Vortices".

INTRODUCTION

To all you good friends I will never meet, welcome to the end of the concept of time.

What you are about to discover is knowledge unlike anything anyone has ever encountered before. It creates one of those rare events, a paradigm shift in human understanding. All science, technology, philosophy, and religion will be forever changed; these changes will radically affect every government, culture, and people of every country in the world. Nothing will ever be the same again.

Although mental revolutions are not as destructive as physical revolutions fought with guns and knives, they create the same results. They begin new chapters in the history of mankind: a new chapter begins today with the publication of this new and revolutionary knowledge.

Since this knowledge affects the lives of everyone in the world, it becomes necessary to explain not only what was discovered, but how and why it was discovered too. Therefore, I have recounted to the best of my ability the curious chain of events that set me upon a lifelong quest in search of the ultimate mysteries of science and religion. And as I take pen in hand to tell this story, it is only appropriate to begin by saying I never intended to discover anything. And yet, I must also state the incredible scientific discoveries contained within this book are not the result of an accident either, I had to find them – to confirm my new religious beliefs!

I make this unusual statement because a number of strange incidents in my life compelled me - forced me to go in search of the ultimate mysteries of the universe. Although when these incidents occurred, I did not recognize them as turning points in my life, they changed me, motivated me, and led me to the wonderful discoveries you are about to encounter for yourself.

Russell Moon, 2019

PART I

Chapter 1
A Strange Story in Readers Digest

I never intended to discover anything. If anybody had ever told me that one day I would undertake a lifelong quest in search of the ultimate secrets of the universe, I would not have believed them. I certainly would not have believed them if they had told me I would eventually succeed and actually discover the knowledge sought from the beginning of recorded history by the world's greatest philosophers, sages, and scientists. And yet, it is exactly what happened.

It happened because of a mysterious incident that occurred one afternoon many years ago in a small country store in Northern Florida. I use the word "mysterious" because over the years I have come to believe what happened on that day was meant to happen. That my arrival at the store on that day at that precise moment was not just a coincidence. It was as if destiny had placed the right man, at the right place, at exactly the right moment in human history for whatever consequences await us all.

It all began in the fall of 1974. The energy crisis sweeping across the country hit South Florida like a hurricane and the construction industry suffered greatly. Its timing could not have been worse.

I had just started working for myself in Fort Lauderdale as a sub-contractor in the swimming pool construction business. But when this mini recession began, people had trouble getting loans to buy pools and I soon found there was very little work here or in any of the surrounding towns. Since I had spent most of my savings getting my business started and no money was coming in, I was going broke fast. So, after carefully considering my options, I decided to leave and look for work elsewhere. For some reason, I had always wanted to go to a town in Northern Florida called Jacksonville. So, I packed my bags and went.

I drove the length of the state, and soon found a tiny apartment on Jacksonville Beach. I got a job on the night shift at a small fiberglass plant, which allowed me to pay my bills and to look for pool construction work during the daytime. However, after personally visiting all of the swimming pool companies in Jacksonville, I learned pool construction was as slow here as it was in Fort Lauderdale. But that didn't stop me.

Undaunted, I next went to all of the surrounding towns looking for work. And when I didn't find any, I tried other towns in other counties too. But no matter how hard I tried; I ended up with the same discouraging results. In one last desperate effort, I drove all the way to Tallahassee – the State Capitol and spent the day contacting most of the swimming pool construction companies in the area. But it was not meant to be. The construction business was the same everywhere; there was no work available, period.

To say I was disappointed was an understatement. I felt like a fool for having come up here. And that night, as I sat in my hotel room, I wondered why I had ever come to Northern Florida.

The next day I found out.

As I was driving back to Jacksonville, for some unknown reason I decided to abandon the road I had driven the day before and take a more leisurely route through the countryside. Perhaps it was just an impulse on my part, or as others have since suggested – perhaps it was all part of destiny's plan. But for whatever reason, when I was about halfway to Jacksonville I turned off of the highway and began to drive along one of those old narrow country roads crisscrossing Northern Florida.

Even though this outdated collection of farm roads ran roughly parallel to the direction I wanted to go, after several hours of frustrating driving behind slower moving farm vehicles I began to wonder if my impulse was a wise one.

Thinking perhaps I was foolish to have taken these slower roads when I might have already gotten to Jacksonville, the sight of a small country store up ahead made me want to stop, get a soda, and take a short break. And that decision changed my life forever.

Just like a plot from a "Twilight Zone" episode, a meeting was about to take place in that little store that will one day effect the lives of everyone upon this planet. A statement which is even harder to believe because nothing about this old, dilapidated building seemed important, in fact, just the opposite.

The shingles were coming off of the roof, the long wooden porch running the length of the building was swayed inward, and the small gravel parking lot had deep ruts in it making parking difficult.

When I climbed the front steps they seemed to creak in protest, and when I opened the screen door, the rusty hinges screeched so loudly I tried to close it as gently as possible to spare the ears of those inside.

But the noise didn't matter. Except for a lone clerk who appeared to be intensely reading an article in a small magazine and didn't even look up when I walked in, the store was empty of people. Empty of people yet crammed full of everything else. Crowded into it was anything anybody ever needed. Packed onto the shelves, leaning up against the walls, and hanging from the ceiling were all sorts of farm tools, sacks of feed, work clothes, cans of food, soft drinks, and yellow straw hats. Everything was there, including that which I least expected to find – the keys to unlocking the secrets to the mysteries of the universe. (Quite a place now that I think about it.)

Although many years have elapsed since that incident occurred, perhaps its continued clarity is due to the fact I have never ceased to marvel at the incredible synchronicity of the events that happened on that day. Wondering over and over again what might have happened if I had not taken those particular roads. Or if I had left Tallahassee just five minutes sooner or later, or if there had been no slow moving traffic to delay my arrival at the store until precisely the exact time needed. Repeatedly I have gone through all of the "what ifs" that might have occurred which could have prevented me from walking up to the front of the store and arriving at the cash register, at the exact moment the young man finished reading the article.

This young man whom I never saw before and never saw again, looked up at me as if seeing me for the first time. He started exclaiming, "I just don't believe it. I just don't believe it."

Then totally ignoring the money in my hand, he started telling me about the bizarre story he had just read in the *Readers Digest Magazine,* which lay on the counter before us. For no apparent reason at all he started talking to me as if I was an old friend, showing me the article the magazine was opened up to.

Although I listened politely to him, my mind was on getting home. Even though I eventually told him I was going to have to be leaving, he continued to talk to me. He kept talking to me until I decided to leave. He kept talking to me all the way to the door. And as I started to walk out, much to my surprise, he grabbed my arm and insisted I take his copy of *Readers Digest* with me. In fact, he almost wouldn't let me out of the door until I accepted the magazine. (I left so quickly I forgot to pay him for the soda, and he forgot to ask me for the money.)

Perhaps he longed to talk to me because I was the only person in the place, or, as mentioned before, perhaps it was all part of destiny's plan. But whatever the reason, I was sufficiently impressed by this man's sincere amazement to read the article myself.

Here it should also be mentioned that even though *Readers Digest* is an excellent magazine, I hardly ever read any magazines at all. I never would have read this magazine and this particular article if it weren't for this man's insistence it was one of the most incredible stories he had ever read. And he was right.

It was one of the most incredible stories I had ever read too - which is an understatement, for it turned my life upside down. Nothing was ever the same again.

This first of its kind article told about what has since come to be known as a "Near Death Experience" of a man from New York City. It seems during the course of a heart attack suffered by this man, his conscious mind was not unconscious at all, as one would normally expect. Instead, he found his consciousness to be in another location away from his physical body; a strange location - where he found himself confronted by a grid-like barrier of light - unlike anything he had ever experienced before. But even stranger was the curious fact that when he was revived, he somehow felt disassociated from his physical body. He felt he was more like an observer, watching what his physical body was doing, rather than being the master of it.

Today, these near-death experiences are reported so often they are no longer considered to be unusual or even unique, but, not in 1974. At that time the general public had no knowledge of these incidents whatsoever. So, to read about one of them for the first time was a wonderful - yet puzzling experience. For even though the people whom I showed the article to were delighted by the spiritual implications of consciousness surviving death, at the same time they were baffled and incapable of explaining how such a strange phenomenon could occur: a paranormal phenomenon, which directly contradicted the theory of consciousness as proposed by western medical science. The theory that consciousness is a product of the neuron activity in the brain, and as such, is limited to the cavity of the skull.

Even though I was just as baffled as everyone else, what absolutely fascinated me was the fact that somehow there was a physical change in the location of the perspective through which this individual viewed the world. In other words, this person's conscious perspective of reality changed from the normal position within the body (just behind the eyes) to an abnormal position outside of it - something which is physically impossible according to our present knowledge of anatomy.

"However, the impossible cannot occur, only the possible can occur". Therefore, when the seemingly impossible does occur, it is the signal man's understanding of himself is either flawed or incomplete. That a new body of knowledge lies waiting to be discovered, which will revolutionize the way man views himself and his place in the universe.

I must admit, even though at that time in my life I was only vaguely aware of this truism, I was suddenly aware of a deeply felt need to understand, to know just how such an event could occur which directly contradicted everything I believed in. Even now it is hard to explain this feeling, but so great, so overwhelming was this need to understand, I was unable to think about anything else. For weeks, until I snapped out of it, I considered, rejected, reconsidered, then rejected many exotic electro-biological and psychological explanations.

For example, I thought it might be possible to create an upward extension of the brain's electromagnetic field away from the body. Or then again, perhaps somehow the unconscious mind of this individual was still active, still alert, and later fooled his conscious mind that wasn't functioning. That he was only imagining everything that happened. Again and again, I tried to come up with a physical explanation until exhausted – I was at last forced to accept the possibility such a phenomenon could also be explained by the existence of a soul within the human body.

That perhaps the change in the conscious perspective of this individual was caused by a premature release of the soul from the body when the body was in a condition that simulated death. That it was not the consciousness of the physical body that was observing what was occurring, but rather the consciousness of the soul. And that when the physical body was revived, and the soul was somehow drawn back inside, "its" memory of "its" experience was fed into the consciousness of the physical body which now perceived this experience to somehow have been its own.

This hypothesis was a major event in my life. I had been raised as an atheist, and had no religious beliefs whatsoever. I did not believe in a soul, I did not believe in God, nor did I even consider such beliefs to be worthy of any serious contemplation or speculation.

But now I could no longer ignore what this incident implied. Especially after being trained to use the reasoning process of the "scientific method" – which has as one of its cornerstone postulates the "uncompromisable", unyielding axiom that all possible solutions to a problem must be considered no matter how strange or implausible they may seem to be to the investigator. Looking back in retrospect, it is kind of odd, but it was due to the reasoning process of the scientific method I was now confronted with the real possibility the human soul actually existed.

Suddenly, I didn't know what to think anymore.

I was a 25-year-old man who grew up believing a man was nothing more than a biological accident. He was born, lived, died, and that's all there was. There was nothing more, and those who believed so were only fooling themselves.

It was a tough time; but then it never is an easy time when an individual is rudely confronted with the unsettling probability that his beliefs are a sham, and that his most basic and fundamental thought processes themselves are built upon false premises.

Yet sometimes, what is originally perceived to be a disadvantage can actually be an advantage in disguise.

Sometimes adversity can transform itself into a powerful motivational force, creating an unrelenting mental drive that lifts and propels an individual above and beyond the confines of his former unfortunate predicament. A mental springboard whose impetus gives him a tireless self-motivation which drives him and forces him to achieve much more than he ever could have accomplished when confronted with the mere presence of an ordinary set of circumstances alone. And so, it was with me.

My lack of knowledge of the subject of the soul, coupled with an intense need to understand everything about it, and hence fill some sort of emptiness within me, now became the motivating and driving force that compelled me to find out everything I possibly could about it. And I did. I did until a problem developed.

Much to my distress, a terrible emotional conflict began to grow within me. This conflict started out small and then began to grow larger and larger until it consumed my life. It set me upon the lifelong quest that culminated "ironically" in the greatest scientific discovery ever made. I say "ironically", because the conflict within me did not deal with science at all. Strangely enough, it centered itself upon one of the great philosophical mysteries of religion: the location of the Kingdom of God. If the Kingdom of God really exists – WHERE IS IT?

Chapter 2
Where Is the Kingdom of God Located?

The terrible conflict seething within me was a conflict many of you may have suffered yourselves. It was a conflict between the vision of the universe containing God, Souls, and the Kingdom of God, and the vision of the universe we see through our mightiest telescopes. A conflict between what can be called the "Religious Vision of the Universe" and the "Scientific Vision of the Universe". A conflict created by the following problem:

When the body dies, and the soul leaves the location of the physical body, it must travel to another location. Or rather, when the soul leaves the co-ordinates occupied by the physical body it once inhabited, it must travel to another set of physical co-ordinates. A set of co-ordinates designating a physical location which has to be *just as real* as the physical location occupied by the soul when the soul was in the physical body.

Since it is assumed this place is a Kingdom of Souls called various names in various religions, one characteristic common to all of these different beliefs is the fact that whatever you choose to call it, this kingdom has to be a real place existing in a real physical location. If not, this Kingdom of Souls is not a real place and doesn't exist. If it doesn't exist, there is a good possibility that neither the soul nor God exists either. And if there are no souls, and if there is no God, religion is a lie!

So where is the location of the Kingdom of God?

This problem began to trouble me, which seems odd, because the mystery of the invisibility of the soul didn't seem to bother me at all then (although it did later). At this time in my life I believed in the tentative hypothesis the soul might exist as some form of "pure energy." Hence, the soul's invisibility would just be a natural consequence of its construction, and this rationalization seemed satisfactory.

But the mystery of the location of the Kingdom of Souls could not be rationalized away. So, much to my own chagrin, every time I found myself wanting to believe in the soul and in God, the mystery of the location of the Kingdom of Souls was always in the way.

Soon, the mystery of the location of the Kingdom of God began to occupy all of my conscious thought. All day long I would find myself thinking about it. The logic of the soul relocating to another position after the death of the physical body was irrefutable. No matter how hard I tried to rationalize it away, I couldn't do it. Nor could I ignore it.

To ignore it meant retreating into a fantasy world where I rejected the great discoveries of Astronomy, something I refused to do. All my life I had been fascinated by Astronomy and the latest discoveries of this subject were still fresh in my mind after having taken the latest college course in it before dropping out of school a year and a half before. So, I was left to suffer with this emotional conflict. I suffered for weeks, until one evening while sitting upon a lounge chair on Jacksonville Beach I could take it no more. Although I liked to sit upon the beach every evening and watch the stars of the night sky slowly appear as the sun set behind me, on this particular occasion I ignored everything and took a hard look at myself instead - and I didn't like what I saw.

It was painful to realize I was going nowhere in life. After spending a hitch in the Navy during the turbulent era of the Viet-Nam War, and after a bad relationship with a girl from South Carolina, all I wanted was to have peace of mind. And for the past year, I thought I had found it.

But nothing lasts forever. The article, my knowledge of the scientific method, and its implications left me deeply troubled. I knew I was going to have to do something about it.

To alleviate the anxiety I was suffering from, I realized I had to start taking some sort of action. So, on this particular evening, I deviated from the normal routine. I carried the lounge chair out away from the buildings and the other people. I hauled it all the way across the beach and positioned it down by the water's edge. And as the sun set in the west and the stars began to appear in the east, I decided to begin my quest for answers by defining the exact nature of the problem bothering me. A problem that begins and ends with credibility. For example:

The credibility of any religious belief that includes a kingdom of souls balances precariously upon a believable location for this Kingdom of God, or Heaven. Never before in the history of mankind has the knowledge of the location of the Kingdom of God been more important. Before the Seventeenth Century, this question was not a problem. Since early religions worshipped the Sun, the Moon, and the planets, Heaven was always assumed to be somewhere up in the sky beyond the stars. Since nobody could see beyond the stars, nobody knew what lay beyond them. Hence, this explanation was satisfactory. Because nobody knew what was right, nobody could say what was wrong. Then came the invention of the telescope, and everything changed.

Four hundred years ago, Galileo turned his newly built telescope into the night sky and saw many wonderful sights, but he didn't see the Kingdom of God. This event probably surprised him as well as everyone else of his era - an unforeseen development that eventually presented the leaders of Christianity with a problem. Since everybody knew "Heaven" is above, and now that we have the means to see it, why can't we see it?

Although the priests of later eras were probably bewildered by this continued failure, the atheists were not. They possessed this powerful argument: *"If it is known to those who believe in Heaven that Heaven is above us somewhere up in the sky, and we have looked for it and cannot see it, it must not be there. Conclusion: The Kingdom of God is not real, it doesn't exist."*

In other words, since the *assumed* location of the Kingdom of God doesn't exist, it was concluded the Kingdom of God doesn't exist either. A convincing argument, even though it is based upon false logic. But false logic or not, since there is no other knowledge to contradict it, it has become an "acceptable solution" to an unwanted problem. Even though it is based upon a lack of proof, those who use it to justify their refusal to investigate further - accept it as a proof.

The reason why this argument is false is simple. To be able to say something exists is easy, all one needs to have is evidence of its existence. But before one can say something does *not* exist, one must know everything that does exist everywhere within the universe. The statement "something does not exist" is provable only by elimination. Elimination that can only be verified by a complete and thorough search of the entire universe. (Note, only after man explored the entire world was he finally able to state that dinosaurs were extinct, and even then, the prehistoric Coelacanth, a fish thought to be extinct for over 70 million years, was discovered in 1938 living in the ocean waters off the coast of Southern Africa.)

Consequently, before any man can state unequivocally, and with absolute certainty, the Kingdom of Heaven does not exist, he must first explore the entire cosmos before such a statement is valid. Without an exploration, such a statement is an erroneous assumption, an illogical opinion. An irrational opinion when one continues to believe it in spite of being made aware of the fact that it is illogical. But it gets worse.

There is another problem that contradicts the above logic. Amazingly enough, the discoveries of 20th Century astronomy now reveal a high probability the Kingdom of God *does not* reside within the physical universe observable through our telescopes.

This little known possibility must be revealed due to the fact that today, many well educated and highly technically trained people of all different religions are trying to integrate their beliefs with the common beliefs of present day astronomy. Unfortunately, many people have reached the equally invalid conclusion that the physical matter of the universe might block the telescopic vision of the Kingdom of God. Or rather, we might not be able to see the Kingdom of God because something might be in its way, like a dust cloud, or another Galaxy, etc.

Although this argument sounds logical, it is probably wrong due to two reasons: the impermanence of the physical universe - as seen from man's observations; and the permanence of the Kingdom of God - as per the words of Jesus in the NEW TESTAMENT.

The impermanence of the physical universe is now one of the cornerstones of Astronomy. Although we see the same sky Aristotle saw, it is not the permanent place he believed it to be. Everything is in motion. Stars are now known to have life cycles. Some explode into Supernova, and some, like our Sun, will temporarily expand outward. Someday, hopefully far in the future, our Sun will begin to expand until it reaches a tremendous size, becoming a Red Giant millions of miles in diameter, vaporizing Mercury and Venus, while boiling away the Earth's atmosphere and oceans. Turning the Earth into a blackened rock before the cycle is completed and the Sun collapses inward upon itself, becoming a tiny White Dwarf star.

Our Milky Way Galaxy is also in motion. Spinning like a gigantic pinwheel, this massive collection of stars moves through the universe with its companion Galaxies called the Local Group. Its fate is unknown. We know Galaxies collide, and we know Galaxies explode. Will ours? Some Galaxies also appear to have giant Black Holes at their centers, which are gobbling up their stars. Is this any place for the Kingdom of God?

Also, the universe appears to be the result of a cataclysmic explosion called the Big Bang. This belief in the "Big Bang" is backed up by physical evidence. The "residue" of the explosion is seen in the Cosmic Background Radiation, and the observation that all of the Galaxies possess a Red Shift, indicating they are all rushing away from each other.

Furthermore, if there is enough mass in the physical universe, it will end in what has commonly come to be known as the "Big Crunch", to disappear forever, or to begin the cycle all over again in response to the "Oscillating Universe Theory". A situation completely at odds with the words of Jesus, which states the Kingdom of God, will be forever. Which is the primary reason why the Kingdom of God probably does not exist within the physical universe; for how can something be forever amidst this maelstrom of change? It is unlikely. Therefore, it is also unlikely that the Kingdom of God resides within the confines of our physical universe.

This conclusion creates a dilemma. For even though the statement, "The Kingdom of God does not exist" is illogical, and even though it is unlikely that the Kingdom of God exists within the confines of the three dimensional physical universe – what is left? Is there any other place to look?

To one raised in a church environment and trained to believe in Jesus from childhood, such a question does not need to be answered. But to a former atheist raised in a scientific environment, it cannot go unanswered.

The mystery of the location of the Kingdom of God does not have to be solved by the individual who is "trained" to believe in God, because through the generations old process of memorization and recitation, he is literally programmed to believe what others tell him to believe. He is taught (by those who teach him), that those who teach him are teaching him "God's Will", and for him to then question them about where God exists, or where the Kingdom of God exists is to really question his own faith. For him to even ask such a question of them reveals there is something wrong with his beliefs. (Which is a very clever answer for those who do not have a clue themselves about how to answer such a question, and don't want to suffer the embarrassment and loss of credibility - and the challenge to their authority by being unable to answer it.)

Nor does the mystery of the location of the Kingdom of God have to be solved by scientists, because science does not acknowledge the existence of God. Only to a former atheist trained in the principles of the scientific method does it become a problem, a problem that does not go away. A problem that stays with you and eats out your guts until you have to do something about it.

When I first began to work upon this problem, I did not know if it was solvable. However, the more I thought about this ultimate puzzle of puzzles, [this conundrum], the more I began to realize the Kingdom of God's invisibility might be the solution instead of the problem, because if it cannot be seen, there might be a reason why it cannot be seen. We might be looking for it in the wrong place. For if it really exists, yet doesn't exist within the physical universe, it must be located someplace else; it must exist in another location where it can be just as real as we are, yet remain totally invisible to our eyesight.

Is there such a place? I had to find it. The search for this answer now meant much more to me than just finding the solution to a question and satisfying some intellectual curiosity. It was the search for the purpose and meaning of life itself. For if the soul does indeed exist, and the Kingdom of God is real, then we aren't just a bunch of biological microbes living upon a speck of sand in the middle of a cosmic maelstrom, as some have insisted. Instead, it means there is a purpose for being alive.

So, what is this purpose I asked myself, and what is its importance?

The answer I came up with goes like this: if life has meaning, then the purpose in life is to find that meaning, but if life has no meaning, the search for a meaning gives one the purpose in life they would otherwise not have. So, either way, the search for the meaning of life – gives meaning to life.

And as I got up from my lounge chair and stood under the brightening sky proceeding the dawn, I vowed to give meaning to my life. From this day forward I would have a purpose in life, not just a vain purpose in pursuit of an impossible goal, but a real purpose. To actually succeed, to do what no man had ever done before, to solve the mystery of the invisibility of the soul and find the actual location of the Kingdom of God.

Although the above statement might sound like some sort of overconfident "bravadoism", one thing I did not lack was confidence. I had supreme confidence in my abilities. A confidence, born of success.

During my military service in the Navy, I was trained to be an electronics cryptographic technician. I was trained to seek solutions to very complicated problems. A job leaving me skilled in finding problems in some of the most sophisticated electronics equipment ever developed. Hence, I believed in myself, and I knew if I could find the solutions to troubles in very complex electrical circuitry, I could also find the solution to the mysterious location of the Kingdom of God.

But I was also a realist. Even though I was confident and full of spit and vinegar, I knew my ability in solving electronics related problems was a direct result of the knowledge I possessed about electronics. While the problem I was now trying to solve required a completely different type of knowledge. What type of knowledge was necessary to solve this seemingly impossible problem of problems?

There was only one solution I could think of: school. Go back to college. There was no other choice.

So, in December of 1974, I packed my bags and headed west to California.

Chapter 3
The Search for the Kingdom of God Begins

In order to discover the mysterious location of the Kingdom of God, I realized I needed to do two things: I had to make an overall study of both science and religious philosophy, and I knew I must think differently from everyone else.

Although I wanted to learn what others knew, I did not want to become another victim of their mistakes. Therefore, I didn't want to think like they had thought. The reason why I didn't want to think like others was due to the observation that they had not solved this problem for themselves. Hence, there had to be an error in their way of reasoning. An error keeping them from breaking the impasse and making the revolutionary discovery necessary to reveal the truth.

The key to making revolutionary discoveries lies in thinking differently from everybody else. It is not hard to understand that when everyone thinks alike, everyone reasons alike, and everyone comes to the same conclusions. When this erroneous reasoning process is passed on from one generation to the next for countless years, no progress is ever made. It continues unless the cycle is broken.

I knew the only way to break this constantly repeating cycle, and reach new conclusions nobody has ever considered before, was to throw away the ideas of past generations and begin thinking the thoughts nobody has ever thought of before. Consequently, to think the thoughts no one has ever thought before, I knew I must first examine the facts upon which the current "thought system" of beliefs are based and ignore any opinions others might possess. After I had accomplished this goal, I was then going to try to discover if there is an alternative explanation that explains the facts just as well as the explanation currently accepted by everyone else.

A good example of the wrong system of thought people can be drawn to and entrapped in for generations, is the vision of the universe where it was once believed the Earth was located at the center of the cosmos. Even though we know today this idea was wrong, it nevertheless survived for thousands of years.

The reason why it survived for thousands of years was due to the fact that each generation continued to reason like the previous one. Each succeeding generation would observe the same motions of the Sun, Moon, Planets, and Stars seen by all previous generations. After seeing what everybody else previously saw, the reasoning processes they inherited caused them to agree with the explanations provided them by their forbears and ruling elite - that everything appeared to be circling the Earth. This idea seemed to be so easily "re-confirmable" for each new generation that came along, that everybody kept on thinking exactly the same way as everyone else before them. Except for a near-contemporary of Aristotle in Athens named Heracleides, who believed the Earth was rotating upon its axis, very few people in history up to and including the era of Copernicus, dared to challenge the explanations provided them and think differently from everybody else.

This type of stereotyped thinking is what I wanted to avoid. So, when I finally arrived in California, my mind was made up not only on what to study, but how to study - to study the facts and avoid the conclusions made by everybody else.

Since I had already spent one semester at Orange Coast Junior College in Costa Mesa California after mustering out of the Navy a year and a half before, I decided this was the place to return. So, in January of 1975 I enrolled in the spring semester at Orange Coast Junior College in the town where I had grown up. Since I was a Viet-Nam era Veteran the government paid for my education, and soon I became a professional student; I studied and got paid for it.

Unlike other students, because there was no curriculum given for what I wanted to study, and since I hadn't told anyone what I was doing, I decided to start out by taking courses in Religious Philosophy and Science. Although, I already knew no school in the world possessed the knowledge of the location of the Kingdom of God, I was hoping to find some sort of clue which might help me in my quest. Since I didn't know where to look, or even what to look for, these courses seemed like as good a starting point as any.

Remembering how a previous course in Astronomy gave me the insight into the impermanence of the universe, and allowed me to realize the observable universe was no place for the Kingdom of God, I logically reasoned other science courses might give me other insights as well. Therefore, while at Orange Coast College, and then later, after graduating and transferring to California State University at Long Beach, I took courses in Geology, Astronomy, Physics, Chemistry, Archaeology, Anthropology, and Calculus.

Unlike the unmotivated years of my youth, because I now had a purpose to study and a reason to learn what I was studying, much to my surprise, I did very well. I got top marks and was graduated from Orange Coast Junior College with a 4.0 grade point average, which wasn't bad for a kid who barely graduated High School. I also found I liked the academic environment at Cal State Long Beach. I excelled in both Calculus and Physics, finding I had a talent for these subjects. But amidst all this success, I was deeply troubled.

Since I was still caught up in the struggle between the scientific vision of the universe and the religious vision of the universe, my mind continually oscillated back and forth between these two seemingly contrary sets of ideas. So, whenever I could, I supplemented my college education by visiting many of the different religious organizations located throughout California. Since I looked at different religions as one might consider different solutions to a mathematical equation, I held no emotional attachments to any religious belief. Therefore, I studied them all.

I studied Hinduism, Buddhism, Taoism, Confucianism, Zoroastroism, Judaism, Christianity, and Islam. I studied the *Upanishads*, the *Tao Da Chang*, the *Batavia Gita*, the *Old Testament*, the *New Testament*, and the *Koran*. I studied the lives of Buddha, Lao Tse, Confucius, Zoroaster, Moses, Jesus, and Mohammed.

I practiced meditation and visited Monasteries and Ashrams. I conferred with all sorts of Priests, Monks, Rabbis, Swamis, and Theologians from practically every sort of religious discipline there was, but I was not at ease. In spite of everything I was doing both in and out of school, I felt I was making no progress towards accomplishing my goal. All I was doing was passing courses, making good grades, and yet somehow feeling empty because of it. The problem of the location of the Kingdom of God still eluded me. I felt deeply frustrated knowing I was no closer to the truth than when I started.

Since I was not at ease, I felt I needed a change. I was tired of living in the city. For some reason I decided I would like to go to Northern California, to the Great Redwood Forest. I had always enjoyed these majestic woods located north of San Francisco and just inland from the Pacific Ocean. As a child, my parents used to take me there during their yearly vacations. I remembered a magnificent grove of Redwoods called the "Avenue of the Giants" and treasured the peace I felt while standing in the presence of these wise old trees.

So, recalling what happened when I went to Northern Florida and ended up at the little store, I felt some impulse was moving me again. Therefore, I went to the university library, read the brochures on all of the schools in the California State University system and discovered California State University at Humboldt. A school located in Arcata, a town right in the middle of the beautiful coastal rainforest. Since this seemed to be the school I was looking for, I went back to my lodgings, packed up my bags and headed to Northern California. And just like before, I was right on schedule.

Chapter 4
Higher Dimensional Space

Although California State University at Humboldt is one of the smaller schools in the California State University system, it is one of the most beautiful. Located far up on the Northern California coast in the little town of Arcata, it sits right at the edge of a magnificent forest of giant trees. Called the "Backpackers School" because many students go there to study forestry, I myself temporarily considered studying forestry with the goal of becoming a ranger naturalist, something I always wanted to be as a child.

But it was not meant to be, because it was not the reason why I had come to this school. Fate had other plans for me. I came to this school to take a brand new course just being offered for the first time. A revolutionary physics course containing within its curriculum - the solution to the mysterious location of the Kingdom of God.

Of course, I knew none of this. Before I arrived at the school, I had gone through the catalogue and picked out the courses I intended taking. However, I was still curious about the rest of the courses not in the catalogue, yet being offered during this particular semester. So, as I idly looked through the list handed to me at the administration building, my interest in physics made me note the intriguing name of one particular course I didn't recall seeing in the book: "Space time Physics". Asking about it, I discovered it was a newly developed course being offered for the first time. A course where Einstein's Theory of Relativity would be discussed along with other Physics subjects of current and curious interest. Although this course was not associated with any of the subjects I intended taking, I thought, "What the heck, it's only a two unit course and it won't take up too much of my time, so I might as well try it". So, I enrolled and unknowingly changed my life.

This new course was exactly what I had come to college for. It covered many of science's most intriguing ideas, including speculation about higher dimensional space. The theoretical possibility of the existence of fourth, fifth, and sixth etc. dimensions of space; unseen volumes of space existing at ninety degree angles to our own three dimensional universe: other universes.

I was never more alert. Unfortunately, nothing much was known about higher dimensional space at that time, so nothing much was said. Also, the existence of higher dimensional space was not considered to be a serious subject. Since the construction of a fourth dimension made of space alone conflicted with Einstein's fourth dimension made of both space and time, the existence of all higher dimensions made of space alone was considered to be highly unlikely. Hence this subject matter was only briefly mentioned, then passed by - but not by me.

This was the first indication I had that there might actually be a real location for the Kingdom of God outside of the physical universe. So naturally, I decided I needed to learn more about higher dimensional space: and I did. For the next few weeks, I totally forgot about school. I researched all of the literature I could find in the school library upon this subject, which was a disappointment, because there wasn't very much at all available. However, what I did learn was enough.

Roughly speaking, higher dimensional space is another volume of space located outside of our three dimensional universe. This volume of space has a unique perspective: an observer in three dimensional space cannot see anything existing in higher dimensional space, however, from higher dimensional space, an observer *can see* everything existing in three dimensional space.

I was fascinated. At last, here was the realistic perspective fulfilling one of man's most universal and widely accepted beliefs about the omnipresent nature and ability of God - from the perspective of higher dimensional space, God can see everything and know everything happening within our three dimensional universe.

Higher dimensional space also fits the requirements for the location of the Kingdom of God. From higher dimensional space the Kingdom of God can actually exist in a real physical location. (A real physical location, yet invisible to everyone in three dimensional space.)

Higher dimensional space is also the perfect location for the soul. If the soul is a higher dimensional creature, even though it is a part of us, it too will be invisible. Although it is hard to believe [but true], the soul can be but a millimeter away from the geometric center of the physical body, yet never be seen!

I was ecstatic. Higher dimensional space seemed to fit all of the requirements necessary for the existence of God, the Kingdom of God, and the soul. Or rather, to say I was ecstatic was an understatement. I could not express the emotional euphoria I was feeling. To dream the impossible dream of accomplishing the impossible goal is one thing, but to actually come across a possible solution created an emotional rush that flowed like a river of life through my body. Every cell was supercharged with energy. I was reinvigorated, refreshed after the long journey. I felt whole again, strong again.

It was a great time to be alive. Higher dimensional space had to be the location I was looking for, for if it exists, it suddenly provided an explanation for many of the ideas about religious philosophy which were only assumptions a day before. However, the key words are, "If it exists". For in Einstein's vision of the universe there are only three dimensions of space, and one dimension of time. There is no fourth dimension made of just space alone.

Or is there?

How many times in the past have men rejected an idea because it didn't fit in with the popular beliefs of their era? Looking back at history, how many men two thousand years ago would have believed that the world was not flat, or that the world did not have to be supported by something else? Who in those days would have believed the world was really a sphere hurtling through a dark void of unimaginable size? A void so vast that if it could be compared to the size of an ocean, the Earth wouldn't even be equivalent to the size of a grain of sand. No, I doubt that anybody then would have believed such a tale. And yet it is true.

So, armed with the knowledge the true reality is often more bizarre than the reality we envision, I began to spend all of my time examining the unusual qualities and properties of higher dimensional space. In fact, I spent so much time thinking about higher dimensional space, I had to change all of my courses from "credit" to "noncredit" to keep from destroying my grade point average and flunking out of school.

I must admit I was uneasy knowing I was deliberately neglecting the future a college education could provide me with by spending all of my time on this seemingly frivolous pursuit, but all of this extracurricular activity eventually paid off. It paid in a way I would never, ever have anticipated – with a spectacular insight into the words of Jesus in the Gospel of Luke. An insight unlike anything anybody has ever glimpsed before - a revolutionary insight bridging the gap between science and religion.

It all happened one morning while strolling through the beautiful forest which grows right up to the edge of California State University at Humboldt. For some reason I had decided to take a morning walk. I had walked across the campus, entered the forest, and was following a narrow winding trail weaving in and out beneath the massive trunks of these majestic trees. Although surrounded by the beauty of nature, my mind was elsewhere. As usual I was thinking about higher dimensional space, and on this particular morning I was trying to envision the entrance into and out of higher dimensional space.

The entrance into higher dimensional space is a most unique motion - unlike any other motion in our three dimensional universe. It is difficult to imagine because there is nothing to compare it to. To enter into higher dimensional space, one has to travel at right angles to all three dimensions of space simultaneously. If that doesn't seem to make much sense, the only way to explain this direction is the use of the word "within": within space itself. Or, when matter is used as the reference point, "within" matter itself - in a direction towards the geometric center of matter.

If you seem a little confused, don't feel bad. Since we think and reason in terms of three dimensional space, higher dimensional space is almost impossible to visualize. This problem occurs because of the way higher dimensional space is constructed. Higher dimensional space is a volume of space at right angles to the volume of three dimensional space we inhabit.

A way to envision the relationship between the different dimensions of space is in the following drawings:

Figure 4.1 One dimensional space consists of one line.

Figure 4.2 Two dimensional space is a plane that is at right angles to the line.

Figure 4.3 Three dimensional space is at right angles to the plane.

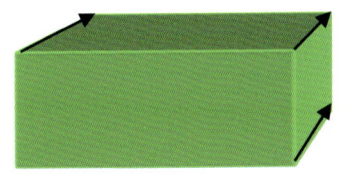

Figure 4.4 Fourth dimensional space is at right angles to three dimensional space. Since fourth-dimensional space is impossible to draw, the following is only a representation, a schematic drawing.

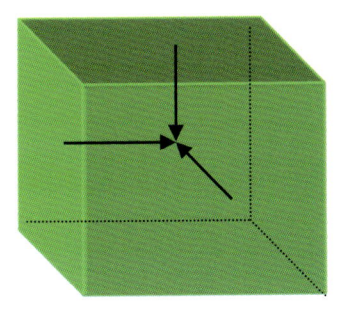

Another way to understand this construction is to look up at a corner of the room you are sitting in. Observe the point where the ceiling and the two walls meet. The ceiling and the two walls represent the three dimensions of space we live in: length, width, and height. Now observe how the ceiling is at right angles to both walls, while both walls are at right angles to the ceiling and the opposite wall. This relationship allows us to observe the fact that each dimension is at right angles to the other two dimensions *simultaneously*.

Although this three dimensional relationship is easy to see, the relationship between these three lower dimensions and higher dimensional space is not. It is easy to say the fourth dimension is at right angles to the first three dimensions, and fifth dimensional space is at right angles to the first four dimensions and so on, but it is impossible to visualize this relationship.

What we can visualize is movement into or out of the fourth dimension. To do this, again look up at the same corner of the room and observe the tiny point where the ceiling and the two walls meet. To reach the fourth dimension we must travel within this point, towards its geometric center, and then outward, into the fourth dimension.

Although this doesn't seem to make too much sense, it does when we realize every invisible point of space within the room you are sitting in is exactly like the point you are looking at. Hence, to reach higher dimensional space, one has to travel within space itself. Or, since every object is made up of an infinite number of such geometric points, a way to describe the location of higher dimensional space to someone who does not comprehend what you are talking about is to use the word "WITHIN".

As a demonstration, you can hold up a small pebble and say to a group of children that the location of higher dimensional space is "within" this very pebble. However, even though this is a dramatic demonstration, it is a poor illustration. Although what you said was true, the children might misunderstand what you are saying and think higher dimensional space is located within this particular rock. So, an even better way to describe the location of higher dimensional space is to use another illustration. For example, you could say higher dimensional space was not only within them, it is also within each and every one of us. By using the matter out of which our bodies are constructed as a reference point, we must travel within - within our own selves: towards the geometric center of our own physical body.

Wait a minute!…What the hell!…The geometric center…Yes! That's it!…The geometric center!

Suddenly I halted in the middle of the trail. Desperately, I found myself trying to recall something Jesus said about the Kingdom of God. Could it be possible? Had my thoughts about higher dimensional space revealed a startling revelation? Could there be a connection between the science of physics and the words of Jesus in the NEW TESTAMENT?

The idea seemed too incredible to be true. I wanted to reject it outright and immediately as nothing more than foolish speculation. Yet I seemed to vaguely recall something Jesus said - a quote in the Gospel of LUKE about the location of the Kingdom of God. For no apparent reason at all, this quote just popped into my mind: just as suddenly, I was running down the trail.

I was never a good runner, but on that day, I was never so fast. I ran like the wind - back through the woods, across the campus, through the streets of Arcata, on the sidewalks and off, dodging in and out of traffic like a crazed wild man until I finally made it the mile and a half back to my lodgings at the old milk factory. With my chest heaving up and down, gasping for breath and the sweat pouring off of me, I finally got the fumbling key into the front door lock, and without bothering to close the door, dashed down the long hallway to my room.

Without stopping I upended and dumped the contents of boxes of books and papers all over the floor in a frantic search for my Bible - which, in my excitement, I forgot was still in the drawer of the nightstand beside the bed. When I remembered, I tore open the drawer, snatched up the Bible and practically ripped out the pages while turning them as rapidly as I could, searching for the quote I was trying to recall. The only quote made by Jesus in the NEW TESTAMENT describing the actual location of the Kingdom of God.

Then suddenly, there it was before me. I stopped turning the pages as my gaze fell upon the red letters of Jesus' quote in Luke 17:20-21. Stunned, I slowly sat down in my little chair. Even though I had read this quote before, I had never understood it. To me, and I know to many others, this quote always seemed to be just another parable referring to the heart or the mind. However, I now knew this wasn't true. Suddenly I knew it meant something more. [Something much more!]

Just how much more I had no idea. Little did I know then, this seemingly insignificant little quote is the philosophical key to understanding how the entire physical universe is constructed. Possessing astounding implications, it is the key to unlocking the great scientific secrets of the universe that have mystified the greatest scientists who have ever lived. Secrets such as the particle and wave theory of matter and energy, the unification of the forces of nature, and many, many more.

Chapter 5
The Greatest Story That's Never Been Told

When I first read the words of Jesus in the NEW TESTAMENT, there were a number of statements and parables I did not understand. But there were also a few which seemed to be so obvious, no other explanation appeared necessary. One of these was Luke 17:20-21. Since Jesus made so many references to the heart and the mind, this statement just seemed to be another parable referring to these human attributes. In fact, I do not ever recall even questioning this explanation. Nor do I know of anyone else who ever questioned it either.

But now, as I sat in my makeshift room at the old milk factory in Arcata, carefully reading and re-reading LUKE 17:20-21, I realized this quote did not refer to the heart or mind at all. I knew then it meant something more, something much more, and oddly enough, I found myself scared.

I was scared knowing I, and I alone was privy to the frightening truth that 2000 years ago, Jesus knew more about the construction of the universe than all of the greatest scientists who have ever lived. I use the words "frightening truth", because at that moment in my life I suddenly knew Jesus possessed knowledge far exceeding anything any man has ever known. How he got it, I did not know. But what I did know was somehow, he was tuned into a source of knowledge that far surpassed the scientific knowledge of any era in human history including ours.

I say, "Tuned into", because he did not get his knowledge from anyone else, "nor" did he get it through experimentation. So, what was his source? Was it God? If it was, then without a doubt, Jesus was the Messiah: the messenger of God, whose coming to Earth was foretold in the *OLD TESTAMENT.*

I was overwhelmed. I felt privileged and honored, yet strangely enough, unworthy. At that most precious moment of all moments, all I could think about was all of the bad things I had done in my life. There were so many others in the world who were much more worthy than I. I was disappointed with my failings; I lay back upon my bed weeping and suffering great anguish.

Sometime during that afternoon, I must have fallen asleep. Later, I awoke in the darkness, realizing I was hungry, yet not caring. Instead, I turned on the light, and opened up the Bible just to look again at what now seemed to be the most important of all the quotes made by Jesus in the NEW TESTAMENT. Like a prospector, who has just discovered a great diamond hidden in the desert, I had to keep looking at it again and again to make sure it was real, and I wanted to be close to it; to touch it; to hold it; to cherish it.

> And when he was demanded of the Pharisees, when the kingdom of God should come, he answered and said, **"THE KINGDOM OF GOD COMETH NOT WITH OBSERVATION: NEITHER SHALL THEY SAY, LO HERE. OR LO THERE. FOR BEHOLD. THE KINGDOM OF GOD IS WITHIN YOU."** (LUKE 17:20-21)

As you read this quote, try to imagine you are seeing it for the first time. I know for many of you who are Christians it is difficult to ignore the training of a lifetime, but don't cheat yourself. Like everyone else who is not a Christian you also have a right to know the truth.

Although most Christian peoples are trained from early childhood to believe this statement is a parable or metaphor referring to the heart or mind, is it just a coincidence it is also an exact description of the location of higher dimensional space? By using physical matter (the physical body) as a reference point, the direction of higher dimensional space (if it exists), is "within", towards the very center of matter itself.

It is also important to understand that if Jesus used another reference point it might cause a conflict. For example, if the physical matter of the Earth, Sun, or Moon was used as a reference, it might create a misunderstanding and lead the listener to believe the Kingdom of God was located within the center of one of these celestial bodies. Or, if he had held up a stone and said the Kingdom of God was within, people might believe the kingdom was within that very piece of rock.

This word "within" is also a very important word. Even though other words are adequate such as "inward", or "inside", the word "within" does an excellent job of representing both the direction and the location of higher dimensional space. However, as before, the three most important words are "if it exists". For in the present view of the universe with only three dimensions of space and one dimension of time, there is just no room for higher dimensional space. Furthermore, ideas and speculations about the existence of higher dimensional space are legacies of the modern era of mathematics. During the time of Jesus, higher dimensional space was unknown. But maybe not to Jesus!

When Luke 17:20-21 is analyzed from the point of view of being a direct answer to a direct question, its implications are too intriguing to ignore. The previous misunderstanding of this quote being a parable referring to the heart or mind is easily understood. Unless one knows something about the geometry of higher dimensional space, this statement appears to be some sort of philosophical reference being made to thoughts or emotions. But this is not so.

To correctly analyze this statement, we must first realize this quote is a direct answer being made to a direct question. Consequently, it is not a philosophical reference being made to the heart or the mind, or as to how one thinks or feels. Nor is this statement a simile or a metaphor.

To correctly understand this statement, it must be remembered the Pharisee's believed the Kingdom of God would be an actual kingdom located upon the Earth. A belief contrary to the teachings of Jesus as demonstrated in John 18:36 **"MY KINGDOM IS NOT OF THIS EARTH"**. Consequently, in response to the Pharisee's question, the first words Jesus responds with, **"THE KINGDOM OF GOD COMETH NOT WITH OBSERVATION,"** are referring to this erroneous belief held by the Pharisees, and Jesus' purpose in answering this question is to correct this error.

Jesus then goes on to say, **"NEITHER SHALL THEY SAY, LO HERE, OR LO THERE."** In other words, the kingdom of God *cannot be seen with, or discovered with one's physical eyesight.* (Note: the word **"LO"** denotes exclamations of discovery.) Finally, he ends this statement by boldly stating, **"FOR BEHOLD, THE KINGDOM OF GOD IS WITHIN YOU."** Although these two statements are short and simple, they are most profound. For if the Kingdom of God cannot be seen with one's physical eyesight, it does not exist within our three dimensional universe. Which means the only place it could exist would be in another dimension of the cosmos.

Because all dimensions are at right angles to each other, the only way to enter the fourth dimension when using matter as a reference point, is to travel at right angles to all three dimensions simultaneously, or "within" - towards the very center of matter itself. The exact direction Jesus is alluding to when he states, **"FOR BEHOLD, THE KINGDOM OF GOD IS WITHIN YOU."** He adds special emphasis to this statement by adding the word, **"BEHOLD"**: which means - "look", "see". A word he seldom used except to emphasize an extraordinary statement - a revelation.

Here we must pause and ask ourselves if we are expecting too much of Jesus. The concepts of higher dimensional space belong to Twentieth Century cosmology, not to ancient religious philosophy.

Or do they?

If Jesus was answering an erroneous question with an accurate view of reality, he already knew about higher dimensional space and didn't have to wait for modern man to invent it.

If this analysis is indeed correct, if it is in fact a correct explanation of Luke 17:20-21, its implications leave us stunned. If the Kingdom of God does indeed exist in higher dimensional space, it means higher dimensional space exists, and if it exists, the present vision of the universe with its three dimensions of space and one dimension of time is in error. Hence, it means that if by using the words of Jesus in the New Testament we can find that error, correct it, and discover the true vision, the course of human history will be altered.

Although Christian peoples already believe Jesus as the Messiah "knew things no man has ever known", for the first time ever, the rest of the peoples of the world will know it too. They will have scientific proof Jesus' knowledge of the construction of the universe exceeded and surpassed the knowledge discovered by all of the greatest scientists who have ever lived.

It must also be emphasized that the knowledge of the existence of higher dimensional space is not just any knowledge. It is very difficult - extremely difficult knowledge to obtain. This knowledge stands upon the apex of a pyramid of other scientific discoveries. It is knowledge obtained only by the most precise and exact experimentation.

Even more important, it will soon become apparent higher dimensional space is not just some mundane, insignificant place. Quite the contrary. Higher dimensional space is the most important place in the cosmos. Without higher dimensional space it will be shown that nothing can exist within our three dimensional universe. Without higher dimensional space, we cannot exist.

In knowing all of this, can there then be any doubt Jesus was the Messiah? Will the attention of the people of the world shift towards this most wonderful man of all men? Can a renaissance in Christianity be created?

The stakes are high. Can it be done?

Can the course of human history be altered by one quote from Jesus?

Yes, it can, if the error in the scientific vision of the universe can be found.

Chapter 6
The Five "Pieces" of the Universe

After discovering the correlation between the words of Jesus in the New Testament and the theory of higher dimensional space, school was never the same again.

I knew if higher dimensional space existed, man's vision of the construction of the universe was wrong. Furthermore, since the error dealt with the construction of space, there also appeared to be an error in the Theory of Relativity - for it was Albert Einstein who developed the current view of space; a deduction which meant trouble.

In the world of science, Albert Einstein has achieved superstar status. An icon of his era, he has a following comparable to that of a cult leader. Even amongst those who have never studied any science whatsoever, the mere mention of the name Einstein conjures up thoughts of great genius and enormous intellect. If science was a religion [and to some it is], Einstein would be one of its greatest saints, and his Theory of Relativity would be one of the favorite chapters in its bible.

Unlike different religions with different sets of beliefs, the same science is believed throughout the world. Almost all scientists, in all cultures and countries accept the Theory of Relativity. Since scientists who believe in Einstein are responsible for training new scientists, nearly all of the scientists in the world end up thinking alike. Consequently, just about every physicist upon this planet sees a mistake in the Theory of Relativity to be a physical impossibility. In every college and university classroom, there is nothing but adulation and praise for Einstein. So unfortunately, within this climate of constant acclaim and approbation for Einstein, for a professor to speculate out loud to his students that Relativity might be wrong, is an act equivalent to the commission of "scientific heresy" in the academic world.

And yet, if higher dimensional space exists, man's current vision of space *is* wrong. Which means Einstein is wrong, the Theory of Relativity is wrong, and Einstein's vision of the universe is wrong. Unfortunately, Einstein's vision of the universe cannot be repaired with a "Band-Aid". The extra dimensions of space cannot just be added to the dimensions already there because there is no sound scientific reason for doing so. Such a change would only be a philosophical one. It would be meaningless because it would not be backed up by scientific proof.

So, I had a problem, I needed proof.

Without proof, I was painfully aware that the existence of higher dimensional space was nothing but supposition and conjecture. However, what gave me hope was the knowledge that if the location of the Kingdom of God really does exist, it will be a provable scientific fact. It will be provable using the techniques of the scientific method: the same method that was used to develop the current scientific vision of the universe. A vision so formidably ingrained into the minds of the present generation that it could only be changed by the same method that created it.

But where do you begin? How do you challenge a theory all of the world's teachers believe in and admire? How do you find a fault in the greatest theory in all of science? Not just any theory, but a theory developed by one of the greatest thinkers who ever lived. A thinker, who may just have been the very best of the very best ever.

Wow. Knowing what I was up against gave me no comfort. It wasn't going to be easy. I also knew I wasn't going to get any help. I was painfully convinced of that, after casually mentioning to a professor of physics I believed something appeared to be wrong with Einstein's construction of space.

If you try it yourself, you will probably experience something like this: first, you will be looked at in disbelief; then suspiciously, like you are crazy or on drugs. Then, after a slight pause, you will be rudely told, "What can an undergraduate student [like you] think they can find wrong with the most difficult and complex theory in all of science - a theory which sometimes takes years to fully comprehend. Above all, how could someone like you, with hardly any training in the science of physics, prove what thousands of the most learned physics professors and researchers in the world can't prove"?

[And most amazing of all, this logic sounds perfectly reasonable.]

I knew it was difficult to oppose such a powerful argument as that one and I realized my prospects of finding something wrong with the Theory of Relativity, and discovering the true vision of the universe didn't look too good. I was in a David versus Goliath battle. But David found a way to win, and while thinking about a way to win, I realized I had one chance too. A slim one, but a real one. I knew from studying history I needed to find some sort of problem with the current scientific vision of the universe. If I could find such a problem, it would indicate the present scientific vision of the universe was inadequate. Being inadequate, it would then become apparent to everyone that it needed to be modified or scrapped altogether.

But could I find such a problem? Where should I begin?

Reflecting back upon the chemistry and physics courses I had already taken; I couldn't seem to come up with anything. But this initial failure didn't discourage me. I knew some sort of problem had to be there - there was no doubt about it. Again, if higher dimensional space did indeed exist, then the present scientific vision of the universe had to be wrong. It was just as plain and simple as that. But what was wrong? Even more important, how could I, a lowly student, discover what has been missed by all of the hundreds of thousands of scientists, professors, and graduate students throughout the academic system of the entire world?

While analyzing this problem, I decided that since I did not even know what to look for, my first task was at least to know where to look. So, I decided to reduce everything in the universe down into its most basic and fundamental parts. I called this "the scientific vision of the universe" - a term used throughout the remainder of this book - and found it consisted of only five things: matter, space, time, energy, and the forces of nature.

Next, I reviewed each part individually looking for any sort of problem.

I started with matter. Matter was easy. The science of chemistry has proven matter is made out of atoms, and atoms are made out of three infinitesimally small spherical particles.

Although hundreds of particles are known to exist, atoms are made out of only three of them: protons, electrons, and neutrons. And it now appears as if protons and neutrons are made up of three even smaller particles called Quarks. [Quarks as discussed in Book 3 of this six part series.]

Figure 6.1

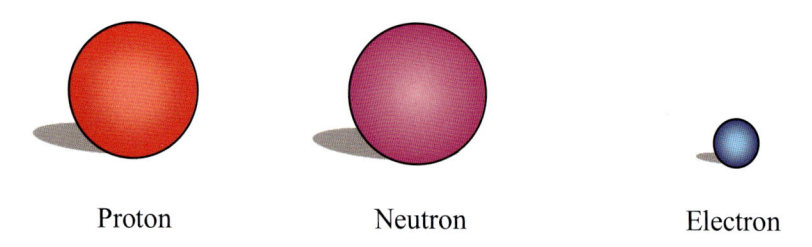

Proton   Neutron   Electron

Protons and neutrons are in the center, or nucleus of the atom, while the electrons are in outer "orbits" or "shells" surrounding the nucleus. The simplest atom, called Hydrogen, has but one proton and one electron.

Figure 6.2

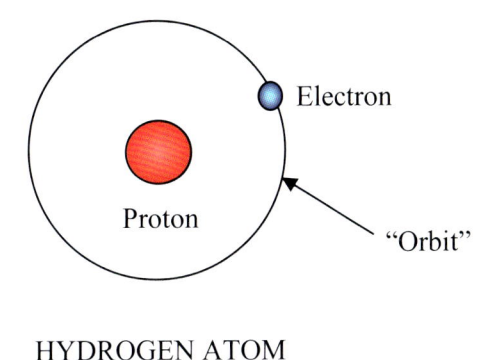

Electron

Proton

"Orbit"

HYDROGEN ATOM

Adding more protons, neutrons, and electrons creates larger more massive atoms:

Figure 6.3

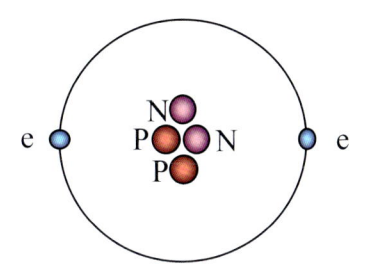

These atoms possess different physical characteristics and are called elements. Currently there are over one hundred different elements. Although some elements, such as gold are usually found in their pure state, other elements like carbon and oxygen combine to form compounds such as carbon dioxide.

Because elements and compounds can be visually seen, I realized I had made the transition from the imperceptible universe to the perceptible universe: a vision I had no problem with. So, I decided to halt my investigations of matter and look at Space. Unfortunately, I failed to grasp the significance of what I had seen while looking at matter.

[It took many years before I was able to comprehend the tremendous importance attached to the spherical shape of protons, electrons, and neutrons. Oddly enough, the spherical shape of subatomic particles is the FIRST of three GREAT CLUES existing in today's mistaken scientific vision of the universe needed to discover the wonderful yet shocking truth about how the universe is really constructed.]

I also missed the problem existing with the 20th Century vision of space.

Here, I knew there was a problem. If higher dimensional space existed, the 20th Century vision of Space *had* to be wrong. But unfortunately, I was unable to figure out what was wrong. All I could say for sure was the scientific vision of the universe sees Space to be made of nothing. However, how can "nothing" be constructed out of three dimensions of space, and a fourth dimension of "time"? This did not seem to make any sense at all.

Scientists are supposed to be very pragmatic individuals. And yet, what they are trying to teach us is that "*nothing is made of something*"?

I knew this contradiction could not be right. But I couldn't do anything about it, because (then) I couldn't come up with any alternative explanation to challenge it. Frustrated by being unable to find the error in something I knew was wrong, I was forced to turn my attention to "Time".

Since Time was considered to be a dimension of space, if higher dimensional space existed, there was probably something wrong with the way this fourth dimension of Time was constructed too. But (then) I couldn't prove this supposition either. Reluctantly, I turned my attention to Energy.

At first, I didn't think I would find anything of interest with Energy. In 1976, the current vision of Energy appeared to be straightforward and simple, or so it seemed. Energy is made up of tiny photons, which possess both particle and wave characteristics.

Thinking about waves, I recalled from physics that even though protons, electrons, and neutrons are particles, they also exhibit wave characteristics too. Because all wave characteristics, such as the waves upon the surface of the ocean, reveal the presence of a medium - the wave characteristics of both matter and energy should be an indication they are moving within some sort of a medium too; that space should be made of something. But I was told this idea was a mistake.

I was told the old Aether Theory, which said space is made of something, was supposedly proven wrong around the turn of the century by what has come to be known as the Michelson Morley Experiment. So, I was again confronted with another "dead end". Because my efforts with energy didn't seem to be getting anyplace, I decided to look at the Forces of Nature.

The seemingly mundane Forces of Nature suddenly become interesting subjects of conversation when they are viewed as the "mysterious glue" which holds the matter of the universe together.

There are four forces: Gravity, the Electro-magnetic force, the Strong force, and the Weak force.

Gravity is the glue holding planets, stars, solar systems, and Galaxies together; it also keeps us from flying off the surface of the spinning Earth. Gravity is supposed to be created by "bent regions of space" surrounding planets and stars, *but again, how can something made of nothing be bent?*

Figure 6.4

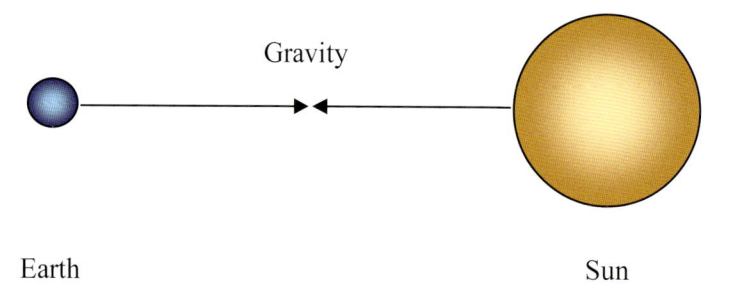

Earth — Gravity — Sun

Anyway, gravity is the attraction of one mass to another mass. Since a sphere's center of mass [center of gravity] is usually located at its center, the center of the earth attracts the center of the sun, and vice versa.

But who doesn't already know this? The knowledge of the workings of gravity is such common knowledge and seems to be so well understood by today's scientists, it appeared to me to be a waste of time to even bother thinking about some sort of error existing here. [But this was a mistake. A great mistake! No scientist living or dead knows the real truth about the force of gravity! The real and undiscovered truth about the force of gravity will shock the world!]

Next, I looked at the Electromagnetic Force. The Electromagnetic Force is the glue holding protons, electrons, elements, and compounds together. It is divided into two parts: the Electrostatic Force, and the Magnetic Force.

The Electrostatic Force is seen in static electricity and lightning; while the Magnetic Force is associated with magnets. These two facts are also so well-known most of us do not give them a second thought. [Perhaps, this is the reason why we do not give two other well-known facts about these forces a second thought either; two vitally important facts that are the SECOND and THIRD GREAT CLUES needed to discover the true vision of the universe.]

Figure 6.5

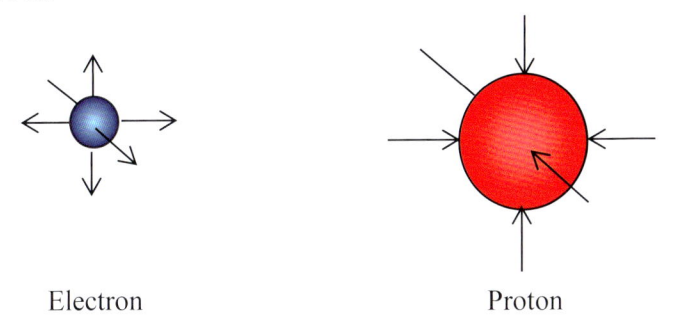

Electron   Proton

[FACT #1 Note how the lines representing the electrostatic force point *directly into or out of the geometric center* of each "particle." Although this simple observation doesn't seem to be of any importance whatsoever, it is the *SECOND GREAT CLUE* to deciphering the true nature of how the universe is constructed.]

[FACT #2 Now take another look at the magnetic force drawn below. Although most of us can remember seeing pictures of the lines of electromagnetic flux surrounding bar magnets, note the curious way they *resemble currents flowing into and out of* the metallic bar. This extremely important observation is the THIRD GREAT CLUE needed to deduce the true vision of the universe.]

Figure 6.6

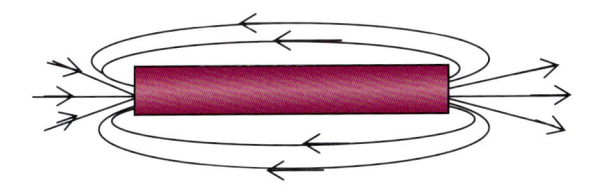

And last but not least, there is the Strong and Weak Forces. These two forces are the glue and "non-glue" associated mainly with the center or nucleus of atoms. The strong force holds protons and neutrons together, while the weak force is responsible for the unusual phenomenon associated with the break-up of neutrons:

Figure 6.7

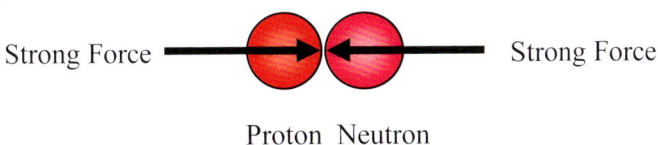

Strong Force ⟶ ●● ⟵ Strong Force

Proton Neutron

[Again, observe how the strong force is somehow concentrated at the center of each particle.]

23

Figure 6.8

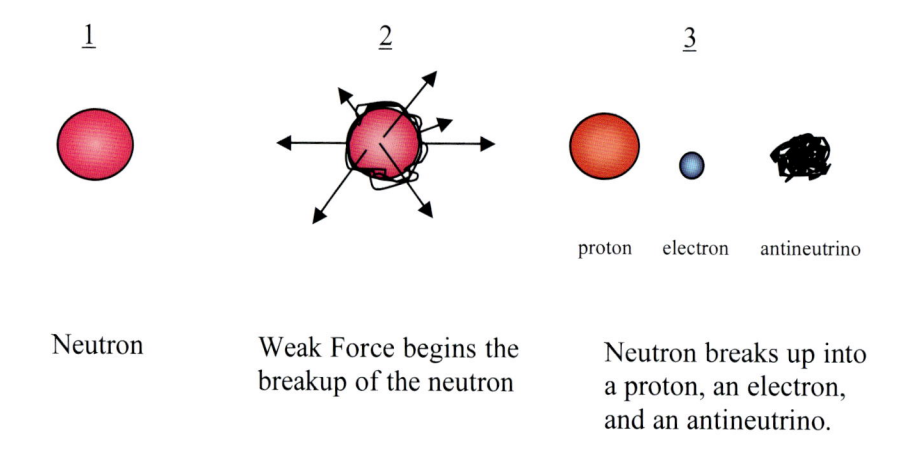

<u>1</u>   <u>2</u>   <u>3</u>

proton   electron   antineutrino

Neutron   Weak Force begins the breakup of the neutron   Neutron breaks up into a proton, an electron, and an antineutrino.

[Note how the breakup of a neutron into a proton, an electron, and an antineutrino almost makes it look as if the neutron is somehow made up of a proton and an electron!!!]

But again, just like matter, space, time, and energy, I could find nothing wrong with the forces of nature. It was terribly frustrating. Everything I looked at seemed to be perfectly normal. But I knew it could not be right. If higher dimensional space really existed, something had to be wrong someplace.

Chapter 7
Hard Times

Identifying a problem is one thing, solving it is another.

Although I wanted to believe higher dimensional space existed, I knew I could not believe in it until I could prove it scientifically. But to prove it scientifically I had to find the mistake in the present scientific vision of the universe. This mistake had to be a major one, a mistake of enormous proportions. But at the moment, this mistake eluded me, and my failure to find it frustrated me to no end.

This frustration created a great deal of stress. The search for the mistake in the scientific vision of the universe was now one of the great quests in my life, but I was also a realist. I knew it might take years of schooling before I was able to discover this mistake, and as mentioned before, I had no illusions about getting any help from Physicists.

It was a tough time. I didn't know what to do. I only knew I wanted to learn the truth, and I now knew I could not learn the truth in school. This attitude made it hard to sit in class anymore.

Being absolutely certain a great error of enormous proportions existed somewhere in the scientific vision of the universe; it was hard to take seriously any of the science courses I was now enrolled in. Nor could I fool myself. If higher dimensional space existed, some, most, or all of what I was being taught was wrong. This realization created a pessimistic attitude I couldn't shake. I became disenchanted with the scientific world. I came to school to find the truth, and now, as I sat in class and looked disappointedly around the room, I became disillusioned. Sadly, I realized many of the people who were here could care less about the truth.

The professor teaching the course was making a paycheck. He was earning a living, teaching others what he had been taught. If he was wrong, what the hell, he still got paid.

Most of the students were also a disappointment. They were here because they had to be here. Many of these science courses were required courses they needed to take to complete their college education and get their diplomas. They were merely doing what they had to, to graduate.

As we used to say in the 60s, "It was a downer." The learning technique was also a disappointment. Students would sit in class writing down, without question or complaint, whatever was said. Then they would go home, study it, and memorize it to be able to pass the tests and get through the course.

What was even more painful was the realization I was no better than the rest. I had fallen into the same trap myself. All I seemed to be doing was sitting in class, taking notes, and passing tests.

I also realized any students who had the nerve to stand up and speak out never got the chance. So much knowledge is thrown at students in such a short period of time, there is no time for the "intellectual digestion processes" to take place. There was hardly any time for introspection and contemplation. The mind was never given the opportunity to sift through all of this new information and look for errors.

Inevitably, I became disenchanted with the academic experience.

I also went kind of crazy.

After the experience of the store in Florida, and the transfer to the University in Arcata, I now believed in the concept of "synchronicity". I believed all a person had to do to succeed, was to hold a goal within their mind and wait to be guided into those particular circumstances where the solution would be found. But, I was impatient.

Although I believe in this mental process today, and know it to be absolutely true, at that time in my life I did not know *how* it worked. All I knew then was that impulsive actions seemed to play a greater role in my life than the carefully thought out ones. Impulsive actions seemed to be the actions that guided me to the circumstances where the solution was waiting to be found. However, in attempting to force this process to occur, I ended up going off on a weird odyssey.

Every time I had an impulse to do something, I did it.

One day I thought I would like to go backpacking. Then I thought that if I wanted to go backpacking, perhaps I needed to. Perhaps at the end of the trail I would find the answer I was looking for. So, I purchased a backpack, bought rice, beans, and a few cans of corned beef, and headed down the road. Just headed down the road not having the slightest idea where I was going or where I would end up. Since circumstance eventually led me to a small town in Northern California called Paradise, I thought this was where my next residence was meant to be. Perhaps this was where the answer I was looking for awaited me? So, after returning for the final few weeks of the semester at Humboldt, I quit school and went to "Paradise".

Because of my religious investigations, I was enchanted with the name "Paradise". An enchantment that soon ended, when to my dismay, I learned the town was originally called "Pair of Dice". A name given to it because of all the riotous gambling which took place there during the 19th Century Gold Rush. So as suddenly as I became interested, I lost interest in that name. However, because this town was located near another town called Chico, which just happened to have another branch of the California State University system located there, I thought this must have been the reason why I ended up in Paradise.

So, I enrolled, I changed my major to Civil Engineering and spent a semester there.

But if I was disenchanted with school before, I was really disenchanted now. I did not know Chico was notorious for being a "party" school where fraternity and sorority groups dominate the town. It was disgusting to see these young adults seeking to fit in with the expectations of their peer groups rather than seeking to develop their own individualities. They were like plastic people being pressed into look-alike molds. Disappointed, I felt another change was needed – this time in me.

Reluctantly, I gave up looking for the hidden meanings of life in every move I made. Dejected, I left Chico and at the end of another long odyssey, finally ended up going back to California State University at Long Beach. But even though I tried to again force myself to study with the same intensity I did when I was there before, it was no good.

The past eventually caught up with me again. Having to take Third Semester Physics as part of my Civil Engineering requirements, I rebelled. Third Semester Physics deals mainly with Einstein's vision of the universe and his Theory of Relativity. Two subjects I knew for sure had to contain major errors.

For weeks I sat in class, discontented. The evidence for the acceptance of the Theory was very impressive. I reluctantly listened as the professor said; "Einstein's brilliant deduction of the fourth dimension of space-time explains many of the unusual phenomenon of the universe. For example, this fourth dimension of time is responsible for the length shrinkage and time dilation effects which occur at near light velocities." (Of course, I did not know then that there is no fourth dimension of "space-time." Even though there is a fourth dimension, a structure within the atom nobody knew existed is responsible for these length shrinkage and time dilation effects).

He then went on to say, "Further proof of the Theory of Relativity is found in the observations of very fast moving subatomic particles. The disintegration of fast moving subatomic particles acts in perfect accordance with the Theory of Relativity. When short lived particles such as the Muon (a heavier version of the electron) are moving faster, they exist for longer periods of time before disintegrating into smaller particles" (I also did not know then there exists a completely different explanation for this phenomenon).

He went on to mention another proof of Relativity was seen in an experiment done with an atomic clock.

An atomic clock launched into space and orbiting the earth at twenty five thousand miles an hour runs slower than a clock located down here with us. And he concluded by saying, "This slowing down of an atomic clock located within a satellite orbiting the Earth is a further confirmation of Mr. Einstein's famous theory." (This statement is totally and completely wrong. This phenomenon is not a confirmation of the Theory of Relativity at all. Later, and again, it will be shown this effect is also being created by unforeseen and unsuspected structures existing within the atom.) However, I did not know then the knowledge within the above parenthesis.

All I knew in 1978 was that when these observations are added to Einstein's use of Relativity to explain some unusual motions in the orbit of the planet Mercury, I myself had to confess this was some pretty impressive evidence. It was extremely hard for me then to deny science's conclusion: "the Theory of Relativity is true, is proven true, and there is nothing more to say." (It was only years later I learned there *is* indeed a lot more to say - much more.)

But in 1978, to say I was discouraged was an understatement. After hearing all of these unbreakable affirmations for which I possessed no alternative explanations, I was deeply troubled. How could all of this evidence possibly be wrong? If only Einstein's fourth dimension of time didn't exist, there would not be a barrier between the three lower dimensions of space and higher dimensions of space. Everything would be fine. But it was there, and it seemed to account for many of the strange time dilation phenomenon observed to exist at near light velocities. The evidence seemed overwhelming. And I was worn out.

I didn't believe Einstein was right and I couldn't prove him wrong. I refused to learn what was being taught. And by refusing to learn I couldn't pass tests.

With this rebellious attitude, no matter how hard I tried to fit in and be a part of the academic system, it didn't work. It was only with great difficulty that I picked up the books I no longer cared about and forced myself to study.

Eventually, I knew I had to go. One day, I just got up and walked out. I stood up, right in the middle of class, and with everybody watching me, gathered up my books, walked across the classroom to the door, opened it, and walked out of the academic world forever, [or so I thought].

I must admit I was sorry to go. Not because I was sorry to leave the academic environment, but rather, because I felt I was a complete failure. I had failed to learn where the error in science existed. I wanted to use the principles of science to prove higher dimensional space existed, but I was unable to do it. And lastly, I believed I had failed to acquire the knowledge or the skills necessary to accomplish either of the above goals. Hence, I was depressed. [Not knowing then, that I had already acquired the knowledge and the mathematical skills necessary to prove the Theory of Relativity was wrong. That another theory about the construction of the universe was waiting to be discovered that would revolutionize science, and prove Jesus' location for the Kingdom of God actually exists.]

But the proof had to wait for many years because I was not well.

Considering the last four years of my life to be an utter waste of time, I had no idea what to do. For a while, I just drifted around. In a kind of "purposeless wandering", I drove all over Northern California and Oregon. I had no idea where to go. Life had no meaning. I did odd jobs to make enough money for gas and food and I slept in my car. When I eventually returned to Southern California to search for a more lasting job, I tried to forget about everything.

I wanted to forget about science, higher dimensional space, and get the problem of the location of the Kingdom of God out of my mind forever. I tried to forget about trying to find a way to prove the existence of anything. I was sorry I had ever even heard about the Theory of Relativity, but the rejection never worked.

Eventually I found myself thinking about the problem all over again, and I wasn't able to concentrate upon anything else. Since I never told anybody what I was doing, most of the people who knew me thought I was crazy. Perhaps I was.

Because of this stress, I couldn't stand to be any one place too long. I would get a job, work for awhile then quit and move on. I was never satisfied with anything I did. I traveled all over the southwestern United States. But eventually, my wanderings found me back in Florida, where, after a number of years, I became a State Licensed Commercial Swimming Pool Contractor. I built pools or worked as a superintendent for other builders.

Without going into details, suffice it to say, many years passed, and many events occurred. But even though I tried desperately to avoid it, every once in a while, I would start thinking about the error in science and my failure to prove the existence of the Kingdom of God. Then, just like a soldier suffering from some sort of Post-Traumatic Stress Syndrome, I would have flashbacks, and suffer terrible feelings of stress and anxiety that sometimes lasted for weeks. While in the interim, I tried my best to cope, to get by and to live with a sense of failure.

And then one afternoon, everything changed. Mediocrity disappeared - exhilaration returned. In a flash of inspiration, I was finally able to deduce the great error which exists in Einstein's Theory of Relativity – the error that allowed me to discover the true vision of the universe; a stunning, spectacular vision no man has ever seen before. A vision of the universe that has allowed me to explain all of the great mysteries of science.

But most important of all, after what seemed like a lifetime of suffering, I was able to scientifically prove Jesus' location for the Kingdom of God actually exists. (And an even more shocking revelation about God!)

It happened like this...

PART II

Chapter 8
Einstein's Mistake

1989 was a good year.

I told myself the stresses of the past were gone, and I was content living a life free of the need to discover the mistake in the scientific vision of the universe. Leave that to others; I was happy in being happy, and wanted nothing else. I was making money, and doing the kind of work I enjoyed doing. I had no problems and wanted none.

Then one day the past caught up with me. Everything changed. In one hour my happy though meaningless little existence was blown away. Because of that day, nothing will ever be the same again for any of us. For on that day the great error in Einstein's vision of the universe was deduced.

It all happened one particularly hot and humid afternoon in Margate, a suburb of Fort Lauderdale. I was working as a swimming pool subcontractor; building homeowner pools for a now defunct company called Swimming Pool World. Working for myself and by myself, I would lay out the shape of the pool, supervise the digging of the hole, form the shape of the pool out of wood, and then finally install the reinforcing bar.

In the winter, the weather was not bad for this kind of work. But in the summer, it got hot and humid down in those holes. And on this particular day, it was hot, real hot.

It was so hot I could only work for a little while in the hole, then I had to climb out, turn on the hose, douse myself with water, and go sit for a while in the shade of a tree.

Like other construction workers, I had no choice but to work in the intense summer heat of Florida. But to beat the effects of the heat, I had developed the technique of intensely concentrating upon a philosophical problem. I found when I did this the resulting introspection effectively removed my conscious mind from its unpleasant surroundings. These mental gymnastics allowed me to keep working where others couldn't. Although in the past I thought about a variety of subjects, on this particular day, I was thinking about "time".

Time has always held a fascination for me. Many years before when looking for the error in Einstein's vision of the universe, I had researched the origins of time. Or rather, I had tried to research the origins of time. Although there are a number of books written about the concept of time, they are incomplete. The concept of time is an ancient concept so old its origins are lost in antiquity.

The early civilizations documenting the concept of time were chronicling a concept created much earlier than the creation of writing. (Note: this idea is not speculation. In 1993, it was reported in *SMITHSONIAN TIMELINES OF THE ANCIENT WORLD* [see OTHER REFERENCES], that a carved bone discovered in the Grotto du Tai in France appeared to be a solar calendar dating back to 10,000 BC.)

The important point to remember about the concept of time is that it was developed during an era of great ignorance. The concept of time was developed during a period when man knew nothing about the motions of the Earth, Moon, Planets, or Sun, which is ironic because these motions are, and always have been associated with the passage of time. For example, the day is one rotation of the Earth upon its axis, the month is approximately one orbit of the Moon around the Earth, and the year is one orbit of the Earth around the Sun.

In fact, since all motion is associated with time, on this particular day, I began to wonder about just what motion it was that brought about man's first perception of this most enduring of all ideas?

Naturally, since both the "day" and the "year" are two important lengths of time associated with celestial motions, I began by telling myself the first perception of time probably began with early man's most rudimentary astronomical observations.

Almost everybody could see day followed night and night followed day. The Sun, Moon, and stars rose above the eastern horizon and set below the western horizon. This entire phenomenon was visible for everyone to observe. However, a very clever individual would have begun to notice an unwavering regularity in these motions and would probably have begun to keep track of them through association with other motions.

Noting the phases of the Moon, he might realize that if he made one mark upon an object for each day of the lunar cycle, he would discover this constantly repeating light show always occurred during the same number of days. He would also observe the night sky and notice how various star patterns would either disappear or reappear during different seasons of the year. And if he were lucky enough to live in one small geographical location year in and year out, he would notice how the location of the rising Sun continually changes its position each day. Slowly trudging northward for six months, stopping, then heading southward for six months, stopping, and then repeating the whole process over and over again, year after unbroken year. Although this particular phenomenon was probably noticed only by later agrarian societies, which tended to live more stationary lives in contrast to the wanderings of the early hunter, gatherer groups.

I was about to continue on with these mental speculations when I was suddenly halted by an astounding idea I never thought before - was dawn age man really sophisticated enough to recognize such an abstract concept as time?

The concept of time is a very abstract idea. It cannot be perceived with the senses. It can only be deduced through observations associated with motion. This observation is extremely important, for many of our conclusions are based upon our perceptions, but sometimes our perceptions are based upon our conceptions.

Consequently, was dawn age man cognizant enough of his mental processes to differentiate between the two? Is anybody? Spoken more bluntly, did he perceive time, or did he conceive time?

When considering this hypothesis, we tend to forget the concept of time is not a concept we as 20th century men developed. It is a mental legacy inherited from early man and passed on from generation to generation. A concept developed within the reasoning process of the child like superstitious minds of our ancient ancestors - men totally ignorant of the astronomical knowledge we take for granted today.

Today, it is hard for many of us to accept the fact that grammar school children know more about the motions of the Earth and the solar system than the greatest philosophers in the Golden age of Greece. It is hard to believe that six and seven-year-old kids know more about the orbits of the planets than Plato, Socrates, or Aristotle. But what is really astounding, is that these men knew far more than their ancient forbears did who were responsible for developing the concept of time.

Since the workings of the solar system are so obvious to us, it is hard for us to envision a time when men knew nothing about the motions of the Earth. It is also easy for us to forget today that ancient peoples did not know the Earth was rotating - creating the effect that everything was circling it. Or that the changing location of the Earth in its orbit around the Sun was responsible for making different star constellations appear at different seasons of the year; or that during its orbit of the Sun, the Earth's 23.5 degree tilt was creating the four seasons.

In fact, everything ancient people were observing in the sky were actually effects being created by motions they were totally unaware of.

It gets even more uncomfortable when we realize our concept of time was developed by men who worshipped fire, feared thunder and lightning as the wrath of the gods, and made animal sacrifices for good crops and successful hunts.

In all honesty, I must admit I was surprised to find myself even thinking these thoughts I had never thought of before. I found myself wondering if anyone else had ever thought like this before? The concept of time is so much a part of our daily lives that to question its validity, even within the privacy of our own thoughts, seems ludicrous - if not insanity itself.

But then, "just because" everyone "knows" time exists, doesn't mean it does. Once upon a time everyone "knew" the world was flat, everyone "knew" the Earth was at the center of the universe, and everyone "knew" Aristotle's vision of an unchanging universe was absolutely correct. Once, to be a member of the educated elite, everyone studied Aristotle and knew the stars were in stationary positions in the night sky. The so-called "educated people" "knew" the positions of the stars were eternal and unchanging. But what is most disconcerting of all, is all of these totally false and erroneous ideas endured from one generation to another for almost two thousand years.

Man has grown comfortable with his belief in time. But suddenly I wasn't. I began to suspect the concept of time might be nothing less than a total and complete mistake, perhaps the greatest blunder mankind has ever made.

Although the concept of time might have started out as a simple concept, in the 20th Century it has become extremely abstract. A situation which causes it to be suspect. I have learned from my analysis of previous erroneous ideas, such as ancient man's belief in the planets orbiting the Earth upon giant wheels, that the more complex the idea becomes with age, the more likely it is to be wrong. This situation occurs because if an idea is wrong, it constantly has to be amended and re-amended as new information is deduced or discovered.

Startled to be even considering the possibility that the idea of time might be a mistake, I began to wonder if anyone in his right mind had ever dared to challenge this most basic and fundamental of all-human beliefs? Is it insanity to think time doesn't exist?

But if time doesn't exist, what does? What were ancient peoples really observing, the effects of time, or the effects of motion? Although motion is associated with time, maybe this idea is wrong, maybe it is the other way around - *maybe it is time that is associated with motion.*

I was about to go on, but it was just too hot.

I overstayed my time limit in the pool, got overheated, and just barely made it up the ladder. I crawled out of the big hole, and half staggered into the welcomed shade of a Banyan tree. The alarmed homeowner, who witnessed my shaky efforts from his window, opened the door and came outside with a glass of water. He asked me if I was all right, and when I didn't answer immediately, he threw it right in my face; apologizing as he did it because he thought I was going to pass out. Which wasn't true at all, I was merely trying to re-focus my thinking onto the problem of time and didn't want to interrupt my train of thought by talking to him.

When he told me he was going down to the store to buy me some Gatorade, I mumbled to him to go. I was glad to see him go. I was glad he was getting out of the way, because after 15 years of aimless searching, I was finally getting someplace. After he left, I observed clouds coming towards me and knew they would bring a welcome drop in the temperature. Knowing I could return to the pool later and do more work when it was cool, I returned to my thoughts; or rather to one single thought which was uppermost in my mind: if "time" was not creating the phenomenon of time, then what was?

It didn't take long to realize there was only one possibility - motion.

Of course! Suddenly it was all so simple. What was early man really observing? Motion. Harmonic motion. And then the truth about the development of time was finally made clear. The development of the concept of time was a mistake. A mistaken attempt made by men ignorant of the motions of the Earth in their efforts to assign a cause to the celestial phenomenon they were witnessing.

I was thunderstruck. The problem with man's vision of the universe was "time". Time itself. "Time" does not exist and never has. The great mistake in Einstein's vision of the universe was suddenly apparent. One of the five fundamental principles of the universe - matter, space, time, energy, and the forces of nature - didn't exist! And Albert Einstein, just like all of his predecessors, had also based his scientific vision of the universe upon something that doesn't exist.

How simple. How extraordinarily simple. I was flabbergasted.

This was one of the great days in my life. On that day, like the flash of light that precedes the explosion, the old vision of the universe was suddenly blown away. The race was on again. The error in mankind's scientific vision of the universe was no longer a possibility, it was a reality. And I was excited.

The universe was not constructed the way we were trained to believe. If I could find out how it was constructed, and if higher dimensional space was part of this new reality, it would prove Jesus knew more about the construction of the universe than all of the greatest scientists who had ever lived. An incomparable discovery of unparalleled importance. A discovery unlike any other ever made in all of human history.

Thinking the above thoughts, I felt a deep sense of responsibility to everyone everywhere. I was in one of the most unique positions experienced by only a few people in all of human history: I and I alone possessed knowledge capable of changing the lives and the destiny of the human race. Knowledge that would directly affect the science, the technology and the religion of all future generations to come.

In honor of the position fate had bestowed upon me, I quit my job and went to work.

Chapter 9
The Curious Relationship Between Time and Motion

In the following months, I spent most of my evenings in the tall, stately library in downtown Fort Lauderdale. It was great to be searching for the truth again, and it felt even greater to have the zest for life return. A feeling I hadn't felt for years. But it had been a long time since I had done any kind of academic study, and I wondered if I was up to the challenge. However, I soon found that old habits die hard, and oddly enough, the construction of this magnificent library helped.

The Fort Lauderdale Library is much more than a library; it is a celebration of libraries. If architecture were music, this place would be a symphony.

I don't recall when I first walked into this library, but I do remember being surprised when I did. I was surprised because instead of seeing the many racks of books I expected to find – a huge gallery five stories high stood silently in front of me. Although there were elevators, one lone staircase somehow found its way up along the western side of this massive and curious void. I use the words "curious void" because every time I climbed this stairway within this building of knowledge, I had the strange sensation death stalked beside me.

I felt this sensation because the stairs were merely slats covered with carpet, and when I looked between them as I climbed upward [especially between the fifth and sixth floors], the lobby seemed dangerously far below. Fascinated by this sight, I realized I was merely stepping upon boards suspended high in the air, and when I did, I knew my mind was wandering and I must quickly re-focus my attention upon the next step. If not, I felt I might lose my perspective, lose my balance, and with just one slip, tumble over the low handrail and spiral to my death far below: a curious feeling within a building whose purpose is devoted to storing the knowledge of the human race.

Perhaps this feeling was intentionally created. Maybe the architect was trying to convey some emotion he was feeling. Perhaps he was trying to tell us all the empty void represents death, while the knowledge we are climbing the stairs to obtain represents life. And if men do not continually focus their concentration upon the next step on the path of knowledge, they might stumble and suffer the ultimate penalty for it. But then again, perhaps I was just daydreaming.

Then again, to dream dreams was why I was there: to dream the dreams no man had ever dreamed before; to think the thoughts no man had ever thought before; and to discover the secrets no man had ever discovered before.

And that is where it happened; in the quiet library standing in the center of the city that had once been the party capital of the United States, I discovered the mysteries of "time".

I would like to be able to say that everything happened dramatically, in one brilliant shining burst of inspiration, but it didn't. Instead, it occurred slowly and patiently, as evening after evening I sat before a table upon the sixth floor and looked out through the giant glass windows at the city far below. Many times, I just sat there silently, watching the day end. Watching the shadows lengthening behind tall buildings and creeping up the sides of the ones behind them. Like a silent chorus of motion, millions of shadows all moving together in harmony. An army of effects, all moving in sync to one grand cause - the Sun. [An observation soon to have great significance.]

Within this special library I was happy. The quiet intellectual atmosphere reminded me of college. I studied there and thought there. And it was there, I first began to realize the inextricable relationship existing between time and motion. An inextricable relationship that begins to become apparent the more one investigates the *measurement* of time.

For instance, I asked myself, is it just a coincidence that all measurements of time are defined by motion? The answer is no. Everything men can think of that might possibly be used to measure the passage of time is in motion. There is no exception. Even tree rings are laid down by chemical reactions taking place at fixed rates.

Chemical reactions also play a major role in "Biological Time". The DNA in plants and animals responsible for the growth and characteristics of all living things are made of chemicals reacting with other chemicals.

Even when we think we are sitting perfectly still, there are billions of chemical reactions taking place within every cell of our body. Some people think they can detect the passage of time by sitting perfectly still in a dark room. But what these people fail to realize is their body is a microcosm of activity. Their hearts are beating. Blood is flowing to every cell throughout their body. Electricity is flowing from their nerves to their brains. Within the brain, neurons are constantly receiving and passing electric current, generating a succession of thoughts. Which all leads us to the realization that even though we are sitting still, our body is a virtual metropolis of chemical and biological motions.

These chemical and biological motions create a sequence of events, which are responsible for our aging process. The chemistry of our DNA becomes a chemical "clock" that makes our body "tick". Knowing this we also know our ages are not a result of time, but the result of chemical reactions.

The realization that no measurement of time is free of motion inspired my imagination. I tried to think of a way to philosophically prove the existence of this unbreakable bond between motion and time. As a result of these efforts, a fascinating philosophical conjecture was conceived. It started like this, "If all of the motions of everything in the universe slowed down, stopped and then slowly started up again, is there any way to tell for how long they were stopped?"

The answer, after much consideration, is no. Because just as a physical yardstick measures the distance between two objects, a regularly reoccurring sequence of events - harmonic motion - is used as a yardstick of time to measure the distance between a random set of events. However, unlike the physical yardstick, when the harmonic motion used to keep track of time stops, the yardstick of time no longer exists and the distance between random events can no longer be measured. Hence, we are unable to tell if the universe was stopped for a second, a day, a year, or a million years.

Also, in such a slowdown, since the motions of everything in the universe would slow down proportionally to the motions of everything else, we would never notice any change occurring at all. We would never notice anything slowing down, stopping, or starting up again. In fact, nothing would ever appear to have happened at all. A simply astounding conclusion.

Based on this conclusion, I began to understand that time is a function of motion, a phenomenon created by motion, and as such cannot exist separately and apart from motion. Just like the shadows I observed every evening while looking out of the window of the library were an effect, time is also an effect and motion is the cause. (Although later, I discovered motion is not the final cause. Motion itself is an effect being created by still another cause. Time is merely the last effect in a chain of causes and effects. But all this will be explained later.)

After coming to the conclusion time is a function of motion, a phenomenon created by motion, something else began to bother me. Each night when I was about to leave the library, I reluctantly looked at the clock to see what time it was. This nightly observation troubled me greatly. Like the rest of us, I had always associated time with clocks, and clocks with time. I was troubled to think a clock, the most common of all mechanical devices, was keeping track of something that didn't exist.

Or rather, I should say I was troubled until I realized that even though time does not exist, the phenomenon of time does. The phenomenon of time is real. Just like the shadows creeping up the sides of the buildings surrounding the library are a real phenomenon, and just like the motion of the stars circling the Earth is a real phenomenon, the phenomenon of time is also a real phenomenon. It exists as a "real illusion".

Since I was excited about what I had discovered, I made the mistake of telling a friend that time exists only as a phenomenon, and not as a fundamental principle of the universe. For my efforts, I was rudely shown a clock and asked, "If time doesn't exist, what the hell is that keeping track of?"

Because of that little incident, I realized I was playing with fire. I sadly understood anytime you attack a belief, you are instigating an irrational emotional response. Therefore, I decided from then on, I would go within, and tell no one what I was doing. But just in case I was ever confronted with the same response again, I researched clocks.

This research revealed the great fallacy of clocks keeping track of "time" is dispelled when one investigates the role of the clock as a navigational device.

In reality, a clock is a mechanical device keeping track of a fixed position upon the surface of the rotating Earth relative to the fixed position of the Sun when it is directly overhead. The hour, the minute, and the second were arbitrary units assigned to time long ago and are meaningless. Over the years, time has merely become associated with clocks and clocks with time.

When someone asks another, "What time is it?" in reality and unbeknownst to themselves, what they are really asking is, "According to the clock - how far has the Earth rotated since the Sun was directly overhead at noon?" Since the circumference of the Earth at the equator is divided into twenty four sections, each about a thousand miles in width, and the clock is divided into twenty four hours, every hour past noon indicates the Earth has rotated a little over a thousand miles. The further breakdown of the clock into minutes and seconds allows this thousand-mile distance to be divided into thirty-six hundred equal parts, which allows the navigator to accurately determine his position to about roughly a third of a mile.

This relationship between the Earthly location of a clock and the position of the Sun is made graphically apparent to modern man by airplane travel. When we travel from America to Europe, we have to change our clocks when we arrive, if not, they will be out of sync with the position of the Sun and their "time" is inaccurate.

The arbitrariness of our increments of time - the hour, minute, second, day, week, month, and year is exposed when considering the rotation of other planets. When men go to Mars, all of these units will have to be changed. Since Mars rotates approximately one half hour longer (24.5 hours) than the Earth's 24 hours, an Earthly clock taken to Mars will soon be out of sync with the appearance of the Sun in the Martian sky. After 12 "Martian days", an Earthly clock taken to Mars will be 6 hours slow. And after 24 Martian days, it will be 12 hours slow. The discrepancy will really become apparent then, because even though the Earthly clock indicates it is midnight, the Sun will be shining high in the sky, showing everyone, it is noon.

If a man upon the surface of Mars tries to navigate using Earthly clocks, he will get lost. He will find the Earthly clock is useless upon Mars. To be able to navigate upon Mars, men will have to build completely new clocks corresponding to the slower rotation of the Martian world.

When he builds new clocks, he will find he will have to change the length of the hour, minute, and second to correspond to the longer rotational period of the Martian planet.

If he stays for generations, the earthy concept of the hour, minute, and second will become meaningless. He may eventually find it necessary to substitute the values of the Earthly second with the values of the slower Martian second. A substitution which will affect his values for all the important measurements which are based upon the value of an Earthly second (such as the speed of light). Hence, all of his mathematical equations for objects in motion will also be affected. He will then find it necessary to have two sets of reference books: one for values based upon the measurement of an Earthly second, and one for the Martian second.

Or he may find it necessary to do away with the concept of the hour, minute and second altogether. He may find it more efficient to divide a Martian day into a "metric system" of units, such as tenths, hundredths, and thousandths.

Although this is all speculation, it is a good illustration to demonstrate our concept of hours, minutes and seconds, are ONLY meaningful upon this planet or upon another planet rotating once every 24 "Earthly hours". When we try to use our clocks upon another planet rotating at a different rate of speed, we find our "earthly clocks" are unsuited for the task. This failure unmasks the truth about the units of time, and we finally begin to realize our concept of the hour, minute, and second is not universal. We begin to realize clocks do not keep track of "time". "Time" is merely associated with clocks, and clocks are merely associated with time.

After writing down these thoughts upon the fallacy of clocks, I decided it was "time" to leave the realm of philosophy and discover a way to prove everything I was thinking. And, lo and behold, after a long search, I rediscovered the Michelson Morley Experiment. [The subject of one of the last lectures I had listened to before quitting school so many years ago.]

Chapter 10
The Michelson Morley Experiment

At the beginning of this book, I made the statement that it seemed as if some of the major events in my life were meant to happen to compel me to go search for the ultimate mysteries of the universe. And now, while thinking about the Michelson Morley Experiment, it made me realize it was almost as if another impulsive action of mine – that took place years before - was meant to happen to keep me from being misinformed.

Years earlier, when I quit school, I thought my rash act was purely impulsive. I was in Third Semester Physics and had just finished studying the Michelson Morley Experiment when I got up and walked out of class. But if I had not walked out at that particular point in the curriculum of that particular course, I never would have been able to discover the error in Einstein's vision of the universe.

Like everybody else who studied Physics, eventually, I probably would have been indoctrinated into believing Einstein's ideas about the fourth dimension of time. Faced with the *supposedly* overwhelming evidence, and the never-ending insistence of today's teachers that Relativity is the truth, the whole truth, and nothing but the truth, it is possible I would have ended up believing Einstein's vision of time was correct. I would never even have thought of questioning the idea of time. Who knows, I might even have accepted the popular belief that anyone who thinks there is something wrong with the Theory of Relativity is either uneducated, or a fool.

Perhaps I might have fallen into the trap trapping so many: human arrogance. I might have believed since I had worked so hard to understand the workings of the most difficult theory in all of science, I was to be congratulated, and perhaps there is nothing wrong with it after all. I might have even tried to incorporate it into the vision of the universe I was trying to deduce, and in the end, would have ended up no better off than any of the physicists who dogmatically teach this theory as doctrine. Luckily, I had avoided that fate.

Although I ardently admired Einstein, I never became a believer. A common trait I unknowingly held with a man recognized as the greatest scientist of his era: Nobel Laureate Albert Michelson.

Michelson was a physicist who became famous for creating what has come to be known as the "Michelson Morley Experiment": one of the most important experiments ever conducted in science.

Tragically, this was an experiment most of the people of the world have never heard about. Which is a shame, considering the explanation of this one scientific experiment alone forces us to define precisely how matter, space, and time are constructed – definitions which are then used to explain how the entire physical universe is constructed. Explanations which both directly and indirectly affect the lives of every person upon this planet.

Although it is hard to believe, the explanation of this one relatively unknown experiment was instrumental in creating mankind's present vision of matter, space, and time. Since the physicists of the Twentieth Century also considered this experiment to be a proof of Einstein's Theory of Relativity (never mind that Michelson rejected Einstein's explanation) a little background information on both how and why it was conducted is essential.

It all began towards the end of the last century, when two prominent American physicists, Albert Michelson, a Nobel Prize winner, and Edward Morley, his colleague, devised an experiment using mirrors and light to detect the presence of the "Aether wind".

Just as a ship moving through the ocean creates a wake, the scientists of that era believed the earth created a similar wake or a "wind" as it moved through the "Aether" - the name given to the substance they thought space was made of.

To detect the presence of the "Aether wind," Michelson and Morley first built a relatively simple apparatus. This consisted of two sets of mirrors set at equal lengths and arranged perpendicular to each other. Between them was a lightly silvered mirror allowing the light beams to penetrate through it as well as be reflected off of it. The basic idea was to turn on a light source, allow a beam of light to first shine on the silvered mirror where it then split - reflected off both sets of the other mirrors, and was then reassembled at a target.

Because the earth is orbiting the sun at an average speed of approximately sixty-six thousand miles an hour, when mirrors three and four (see Figure 10.1) were parallel to the Earth's direction of motion and one and two were perpendicular to it, the mathematics indicated the travel times between mirrors one and two should be shorter than the travel times between mirrors three and four. And the interference pattern created by the different arrival times of these two beams of light at the target should indicate this result, but it was never seen. To Michelson and Morley's surprise, the pattern they saw at the target indicated there was no difference in the travel times whatsoever.

[To simplify the principles involved, the following abbreviated diagram of Michelson and Morley's apparatus is placed against the backdrop of the Earth.]

Figure 10.1

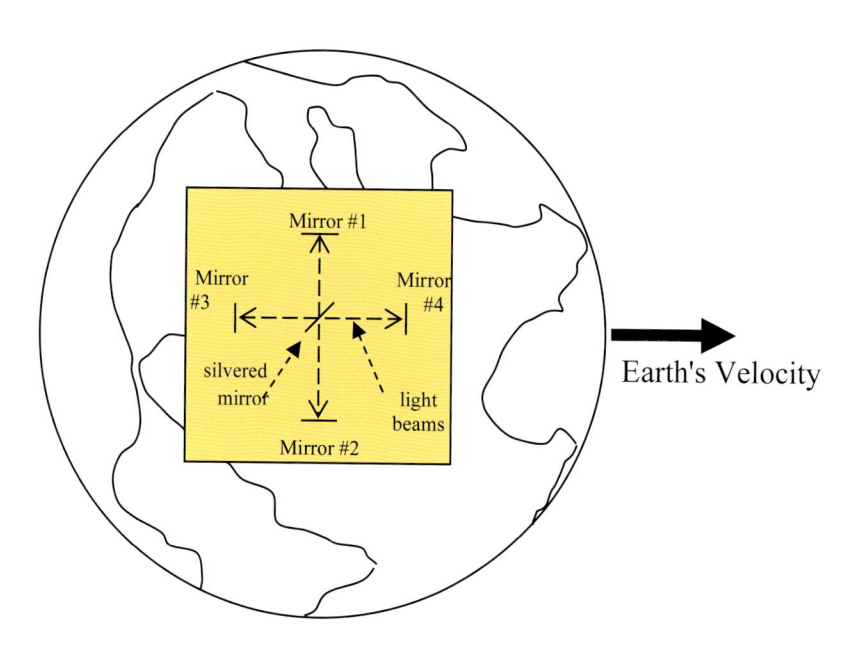

Although surprised, Michelson and Morley were undaunted. They simply built a better apparatus and tried the experiment all over again. But just as before, they still came up with the same results.

Although other people might have quit, Michelson and Morley didn't. They improved the apparatus and tried again; and again and again, but to no avail. The pattern on the screen continued to show the two beams were always traveling at the same speed.

But Michelson and Morley were true believers; they never gave up. They tried for twenty years. They improved their apparatus until it was so sensitive to vibration, it had to be placed upon a two ton slab of sandstone floating in a pool of mercury because an ox pulling a cart down a road close to their building would interfere with it. But even that didn't change the results.

No matter how long they tried, and no matter how sensitive they made their apparatus, the results were always the same - the differences between the travel times they expected to see were not seen. There was no difference between the two travel times whatsoever.

Enter George Fitzgeralds and Hendrik Lorentz.

Fitzgeralds an Irish physicist, and Lorentz a German physicist, both working independently of each other, thought of a way which would make both of these travel times equal. Each one of them proposed if the Aether wind compressed the matter of the earth in the direction of travel by "just the right amount", the distance between mirrors #3 and #4 would shorten. This shorter distance would then shorten the travel time of the light beam moving between these two mirrors, making its travel time equal to the travel time of the beam moving between mirrors #1 and #2.

Note: in the drawing below, the shrinkage in the diameter of the earth is extended for illustration purposes.

Figure 10.2

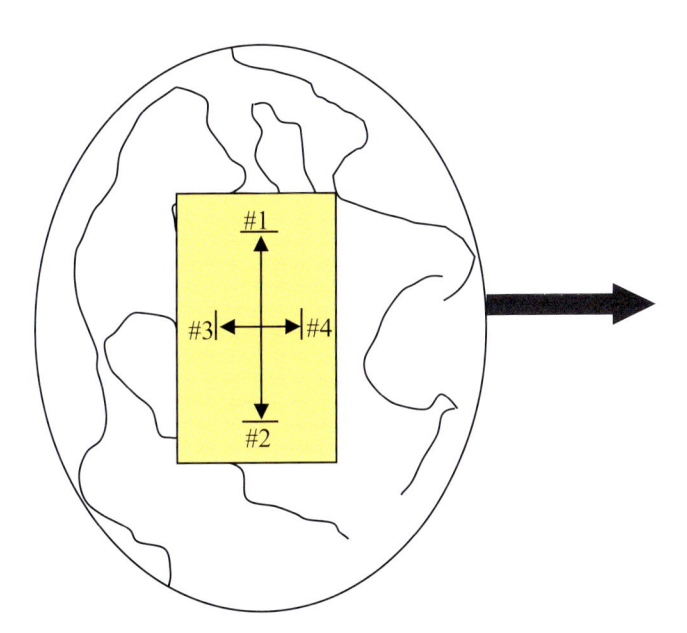

Also, since the Earth speeds up slightly and slows down slightly during the different seasons as it orbits the Sun - although these changes are incredibly minute - because Michelson and Morley's apparatus was now so super sensitive, this change should have been seen too. But it wasn't.

Because it wasn't seen, Lorentz further suggested that if time "itself" somehow slowed down by "just the right amount" no change would ever be noticed. However, the seemingly improbable idea that the Aether wind somehow compressed matter "just enough" to fool the instruments was seen by other rude physicists as an "ad hoc" solution to the problem and was dismissed as being too contrived.

Since no other explanations were available, no other explanations were presented. And the world of physics had to wait impatiently for a solution.

But they didn't have to wait long. When Albert Einstein's ideas began to influence the world of physics, the solution to the problem seemed apparent. According to Einstein's vision of the universe matter was made of something, space was made on nothing, and time was relative and existed as a fourth dimension of the universe he called "space-time." This meant there was no Aether and no Aether wind. This also meant the measurement of time was relative to the speed of the observer. And when these ideas were combined with his brilliant deduction that the measurement of the speed of light was the same for all observers, scientists finally believed they had finally found the explanation for the Michelson Morley experiment.

However, even though Einstein's vision of matter, space, and time has been extremely successful in some areas, it has utterly failed in others. The most important of these failures is its inability to unite the four forces of nature into one "grand unification theory", or adequately explain why they cannot be united: a seemingly trivial technical observation that acquires a status of great philosophical importance. Even those who are uninitiated into the workings of science can easily understand that the true vision of the universe must be able to explain not just some of the phenomenon of nature, *but all of it. Every last bit of it!*

A deduction Einstein himself must have been acutely aware of because he spent most of the remaining years of his life trying to unite the forces of nature into his own personal vision of matter, space, and time. Sadly, failing in the attempt because there was no chance of succeeding. No chance of succeeding because Einstein's failure was not due to a lack of ability or effort, but rather due to his erroneous vision of one of the basic building blocks with which he used to construct *his* vision of the universe – "time."

Because if time is not a fundamental principle of the universe; if time is a function of motion and cannot exist apart from motion, then time is NOT a building block of the universe and cannot be used as one. If time is a function of motion, it is a phenomenon created by motion. Hence, it cannot be used as the fundamental cause to explain time dilation. If it cannot, then the length shrinkage and time dilation effects proposed by Lorentz in response to the results of the Michelson Morley Experiment *are being created by the increased velocity of the atoms themselves as they move through space.*

Consequently, there must exist a hidden, intrinsic relationship between matter, space, and velocity that nobody previously suspected. A secret relationship which unites rather than separates these three seemingly different aspects of the universe's construction. A relationship whose interactions create an exact mechanical and mathematical explanation for the results of the Michelson Morley Experiment - a relationship revealing the *True Vision of the Universe*.

I was excited. Although science presently believes the true vision of the universe was discovered when Einstein introduced the Theory of Relativity, I now knew this was wrong.

The true vision of the universe had not been discovered at all. It was still there, waiting to be found. A sight no man has ever seen before.

Suddenly, I felt like an explorer of old. An adventurer who hears a legend about a lost continent at the end of the world; who gets a ship and sails towards the horizon in search of whatever awaits him.

But this time, it would be no ordinary quest. This would be the ultimate quest for the ultimate prize. The ultimate search for the ultimate vision of everything. And like the view of the Grand Canyon, or Niagara Falls, the ultimate vision of the universe must be an awesome spectacle to behold.

But to be able to be the first man to ever see this true vision of the universe, I realized I and I alone would have to first discover the secret of how matter, space and velocity are related.

Why me? There was no one else. Those who can discover it won't, and the rest never will.

Those scientists who are capable of discovering the answer are unwilling to challenge the Theory of Relativity. They are either true believers in Relativity, and don't think there is anything wrong with it, or they are afraid of the loss of their credibility if they make any controversial statements to their colleagues. So, it is sad to realize those who can discover the truth about the universe - won't.

Unfortunately, the rest of the people in the world will never discover the truth either. They do not have the unrelenting motivation, the "time", or the desperate need to discover the answer to this problem. Therefore, I knew it was up to me.

To me? Suddenly, I felt alone. Who in the hell was I? I was not a physicist.

Feeling inept, I desperately wanted an ally, a seeker like myself. Someone who passionately sought the truth and would go anywhere and do anything necessary to discover it. But I knew no help would be forthcoming. I was alone. It was a hard truth to accept, so I reluctantly accepted it.

I had started alone, and I must now finish alone. I had come much further than I had ever expected. A little bit further wouldn't hurt. The hardest part of any journey is the last part. So, I prepared myself for the finale.

I bought books on math and physics. I reviewed calculus and differential equations. I went to the library and studied the latest discoveries in science. To increase my self-confidence, I lifted weights in the evening, and did pushups and sit-ups in the morning. I didn't crawl out of bed; I leaped out of bed. I fasted, lost weight, and felt good. To be a winner, you must first feel like a winner. To succeed, you must already be a success.

And I was successful beyond my wildest dreams.

I took a mental journey through the innermost foundations of the physical universe. I saw what no man has ever seen before, and I discovered the answers to the greatest mysteries of both science and religion:

Chapter 11
Bent and Flowing Space

Years before, when I was in school, I reduced everything in the physical universe down to its five most basic parts: matter, space, time, energy, and the forces of nature. Next, I tried to define how each piece was constructed, and then I looked at the individual parts searching for some sort of error I knew must exist. But I failed to find it.

I failed to find it because I did not know where to look, or what to look for. But not anymore, now I was ready.

Space was where the problem was. I knew this to be true because if time does not exist as a fundamental principle of the universe, then the fourth dimension made of "space-time" does not exist either.

This error can only mean one thing, that just like the mistakes of our ancient ancestors, 20th Century science has also based its vision of the universe upon the effects it sees instead of the cause of those effects. The explanation of space is based upon what appears to be there instead of what is really there.

So how can this disparity be resolved? How can the true vision of space reveal itself?

The answer is easy. The true vision of the universe might be revealed if some characteristic of space is found which will allow us to change our perspective and see space from its point of view instead of from our point of view.

So where do we look? There is only one place - amidst the current beliefs of space.

According to the present scientific model of the universe given to us by Albert Einstein, space is "believed" to be made of nothing. But even though it is made of nothing, it is somehow "believed" to be constructed out of three observable dimensions of length, width, and height, and one dimension of time. (This doesn't seem to make any sense, because how can something made of nothing have any characteristics at all?)

Furthermore, Einstein also proposed space appeared to be bent around massive objects such as stars, and that bent space is equivalent to the force of gravity. (However, as was said before, how can something made of nothing be bent?)

Although this idea doesn't seem to trouble scientists, it troubled me; because how can those who claim to believe only in those things perceptible to the human senses, turn right around and profess to believe in the "bend-ability" of a substance made of nothing at all? Such logic makes no sense. The only idea that made any sense was for space to indeed be made of something.

Therefore, temporarily accepting this premise, I proceeded along this line of reasoning to see where it would lead me; and lo and behold, I found the bend-ability of space, combined with a 20th Century discovery made by Astronomers, is the first key to understanding how space is really constructed.

After studying Astronomy, I realized the bend-ability of space does indeed appear to be a real phenomenon. Space appears to be bent around massive objects like stars. During solar eclipses, the sky closest to the Sun was dark enough for astronomers to see the stars, the bright light from the Sun usually obscured. When pictures were taken of these stars and compared to other pictures taken at night of this same region of the sky, the stars in the pictures taken during the eclipse appeared to have changed position slightly. An effect indicating the space close to the Sun was bending the path of the light passing through it. An observation that seemed to indicate the space closest to the

Sun must somehow be bent.

Accepting this conclusion as a real observation, a real effect, I decided to investigate its cause.

While thinking about it, I realized my first job was to determine if this bend was inward or outward. Using the force of gravity, this problem was easily solved: since the force of gravity pulls objects towards the Sun, I hypothesized space was bent *inwards* towards the Sun rather than away from it.

Next, because this inward bend was creating real effects, it was only logical to surmise it must be created by something that was just as real. So, what was it?

Even though stars are enormous beyond the ability to describe, I knew a star is nothing more than a massive collection of microscopic individual particles. So, if space was indeed bent around a star, the bend in space had to be a result of all of the individual bends surrounding all the individual particles of which the star was constructed. Since a star is constructed out of atoms, and atoms are constructed out of a nucleus of protons and neutrons surrounded by a swarm of electrons, these particles had to be responsible for generating the bent space surrounding the star.

In other words, the summation of all the bent spaces surrounding all of the particles in a star was somehow creating one large, massive bend in space. But which particles were responsible for the bend? To assume they were all responsible might be a wrong assumption.

In order to solve this dilemma, I took another look at Isaac Newton's law of gravity, noting the force of gravity is directly proportional to the amount of mass present. I reasoned that since the majority of the mass of the atom is contained within the protons and neutrons [which are approximately eighteen hundred times more massive than an electron], the majority of the mass of a star must consist mainly of protons and neutrons.

Consequently, I concluded the majority of the bent space surrounding a star had to be the result of the summation of all the individual bent spaces surrounding protons and neutrons (the electron will be discussed later). Therefore, the bend in space surrounding a star somehow had to be a direct result of the individual characteristics of the protons and neutrons themselves. In other words, protons and neutrons had to be generating the force of gravity.

Although this deduction seemed elementary, try as I might, I could not yet figure how these particles were generating a force. However, the more I thought about the bendability of space, the more I began to realize space must be elastic, and if it was elastic, it had to possess the ability to stretch.

Stretch? Of course. Why hadn't I thought of it before.

Chapter 12
"Less Dense, and Flowing Space"

The ability of anything to stretch is very important.

When something stretches, its density changes. This fact leads to some startling possibilities about space.

If space was indeed bent inward around the Sun, space would be stretched. This stretched space would be less dense nearer to the Sun than further away. Since the Sun is a sphere, a large spherical region of less dense space would surround it, like a pea surrounded by a basketball.

This would mean the bent space theorized to surround planets and stars is only an effect being created by something else. It is a phenomenon being generated by less dense space. Space only appears to be bent in a spherical shape because the Sun is a sphere.

This deduction is important because it indicates the force of gravity would *not* be the result of "bent space", as Einstein believed - *but rather, the result of less dense space.* **And this deduction leads to an even more startling conclusion for the individual particles the sun is made of.**

When the Sun moves through space, the space directly ahead, in the pathway of the Sun, is continually being bent inward towards the Sun as it approaches, then back outwards to its original position as the Sun passes by. Consequently, since stars are mainly composed of protons and neutrons, the same phenomenon would also occur as they too move through space. Making it logical to conclude that space itself ebbs and flows in wave-like patterns as matter passes through it.

This conclusion made me realize if the four forces of nature are in some way related to each other, then, they are all effects being created by the same cause. Therefore, if gravity is the result *of less dense space*, it is possible the other three forces of nature might also be manifestations of space. If space possesses an elasticity allowing it to bend - giving it the ability to move - maybe it can flow too.

If this is indeed so, the electrostatic charges of electrons and protons could be the result of flowing space. Just as less dense space could be responsible for the force of gravity, flowing space might be responsible for the creation of the electromagnetic force. (Remember the THIRD GREAT CLUE from Chapter 6?)

This was an exciting idea, and a fascinating vision to behold. Just imagine, currents of space flowing like miniature rivers through the universe. It was a delightful thought. Was I the first person to ever conceive it? Amidst the billions of people thinking trillions of thoughts, was this thought never thought of before? I knew instinctively it was worthy of closer inspection; but little did I know then this intriguing idea was but the key to deducing the fantastic truth about the particles we call electrons, protons, and neutrons! (They are not particles at all.)

As explained in Chapter 6, the Electromagnetic force is the name given to the positive and negative "poles" associated with magnets, and the positive and negative "charges" on tiny particles such as electrons and protons. This Electromagnetic force is divided into two forces: the magnetic force, and the electrostatic force.

The magnetic force is mainly identified by the *lines* called "flux" which are seen surrounding magnets (Figure 6.6); while the electrostatic force is associated with *lines* pointing directly towards or away from particles like the proton and the electron (Figure 6.5). *Lines* which might really represent "currents" of three dimensional space flowing around magnets, and into and out of protons and electrons.

Now if this vision is real, if these currents really do exist, and if the electromagnetic force is the result of flowing space, there has to be a reason why three dimensional space is flowing. Also, because there is a positive and negative charge, there has to be two different directions to the flow. With one flow moving towards a particle of one charge (such as the proton), while the other flow moves away from a particle of the opposite charge (such as the electron).

Combining these two ideas, I concluded that since the electrostatic charges point directly at protons and electrons from all three directions at once, then three dimensional space must be flowing directly into one particle and directly out of the other. (Remember the SECOND GREAT CLUE from Chapter 6?)

But where? Where is three dimensional space flowing to, and where is it coming from?

There was only one answer: higher dimensional space. Of course. It was suddenly all so simple.

Higher dimensional space would be the exact location for three dimensional space to flow into and out of. In other words, three dimensional space must be flowing directly into one particle, and then out of this particle into higher dimensional space. While from the other particle, the three dimensional space is returning from its trip into higher dimensional space, and flowing back into the three dimensional space it originally came from.

Suddenly, I sat back in my chair stunned. Was it possible?

Yes, it was. If the above reasoning is true, it leads to a simply astounding conclusion - that protons and electrons are not *particles* at all.

Protons and electrons are three dimensional holes in space!

Chapter 13
Protons and Electrons Are Holes in Space!

Three dimensional holes! What an incredible vision.

Just imagine all of the matter in the universe being made of nothing at all. Never, even in our wildest dreams would anybody ever think matter is made of nothing. Like "time", it is one of only a mere handful of beliefs about the universe none of us would even consider challenging.

Our concept of matter is very important to our mental well-being too. Matter not only gives us physical support; it also gives us a certain amount of mental support as well. Without ever thinking about it, we feel secure as we stand upon it, sit upon it, and sleep upon it. If you have ever been upon a ship in a rolling sea, you understand the statement; "It is good to be back on solid ground again."

The idea matter is made of something is perhaps the most unquestionable question there is. Next to the concept of time, it is perhaps the single most commonly shared belief amongst all of the people upon this planet; and to suddenly throw it away temporarily knocks us off our psychological center of gravity.

If you haven't yet felt the full impact of this disturbing idea, look at your body. Its "solidness" is an illusion. It is really a massive collection of microscopic holes. It is shocking to contemplate, but we are made of nothing at all!

I have mentioned these uncomfortable feelings because the deep contemplation of this revelation created a psychological uneasiness within me that was only comforted by my religious beliefs. I found solace from the awful thought "the physical body is a void", in the soothing and comforting belief that we also possess an immortal soul. An immortal soul that God himself created: a second, unseen part of ourselves existing in higher dimensional space.

Feeling a little better, I pressed on, but not without mixed emotions. Although I was intellectually thrilled, I needed to know if three dimensional holes could actually exist. Was there any precedent upon which to justify this truly extraordinary and revolutionary conclusion?

The answer was yes.

From my previous studies of space and its dimensions, I realized that in three dimensional space, there could only be two dimensional holes. All pipe ends, doors, windows, cave openings, etc. are two dimensional holes. Furthermore, the tops and bottoms of tornadoes, hurricanes and whirlpools are two dimensional openings allowing passage into the third dimension.

The mental picture of these two dimensional openings allowing passage into the third dimension is the analogy allowing us to realize all holes are one dimension smaller than the number of dimensions that are present.

This relationship occurs because each surface is one dimension smaller than the volume it envelops. For example, the surface of a three dimensional volume is two-dimensional, while the surface of a fourth dimensional volume is three-dimensional. With these ideas in mind, it is easy to see that to enter into the interior of a volume, its surface must first be penetrated. Because the surface is always one dimension smaller than the volume, the hole is always one dimension smaller than the volume it enters.

Therefore, if fourth dimensional space does indeed exist, even though it cannot be seen, one characteristic indicative of its presence will be the existence of three dimensional holes. Just like two dimensional holes allow passage into the third dimension, three dimensional spherical shaped holes allow passage into the fourth dimension.

If fifth dimensional space exists, there will be additional fourth dimensional holes* <u>within</u> these three dimensional holes, and so on and so forth depending upon how many dimensions exist, and how many dimensions are penetrated.

The incredible vision of space flowing into and out of these three dimensional holes cemented my belief that space had to be made of something. For if the electron and the proton are indeed three dimensional holes in space, and if space is supposedly made of nothing, then how could nothing flow into nothing?

The only logical answer was space could not be made of "nothing". Space had to be made of something. Nor is this idea a return to the old Aether Theory.

In the Aether Theory, it was believed space was made of something, and matter was made of solids. This relationship was similar to the way in which ice floats in the ocean (which is probably where the idea originated). The vision that matter was made of something, and space was made of something caused men to believe the earth displaced the Aether and created a "wind" as it moved through it – in the same way a wake is created in front of a boat. A wind, the Michelson Morley Experiment proved did not exist.

In contrast to the old Aether Theory, this analysis was slowly revealing that contrary to present belief, it was *space that was made of something and matter that was made of nothing*.

Space could now be viewed as a multi-dimensional substance whose three dimensional surface (our three dimensional universe), is inundated with a myriad of three dimensional holes: kind of like Swiss Cheese. Furthermore, there is no "Aether Wind" because matter does not displace space as it moves. When matter is constructed out of holes, space is not "blown" to either side as matter passes by. Instead, space reconfigures itself around these three dimensional holes as they move.

Although it might sound kind of hackneyed to use the analogy of the holes in Swiss cheese, this concept helps envision the relationship between matter and space. The perspective is just reversed. For example, in Swiss cheese, we see the cheese but not the holes, while in the universe, we see the Spherical holes but not the cheese.

It is fascinating to know that if you can see the holes but not the space, these holes will look like tiny spheres - exactly how present day science sees protons and electrons. (Remember the FIRST GREAT CLUE from Chapter 6?)

Philosophizing upon this new relationship between matter and space made me think of the great poetic pathway to God enlightenment in Hinduism: The Jnana Yoga, the pathway of knowledge. However, instead of saying, "I am that, Thou art that, That is that", perhaps it should be said, "I am not that, Thou art not that, and that is not that." because: *"Everything is nothing, and nothing is everything!"*

Although this is a play on words, - it is also a good description of the universe. All of my life, like everyone else, I just naturally assumed I was looking at the picture upon the photograph. Instead, I now realized I was really seeing the reverse image - the negative. It was hard to believe, but everything is opposite to what it appears to be. That which is made of something is not seen, and that which is made of nothing is seen.

And as I continued to toy with these ideas, I finally broke through the mental barrier between the 20th Century vision of the universe and the true vision. And when I did, almost instantly I achieved my first triumphant success. I discovered the answer to one of the great scientific mysteries of the universe. It happened like this…

Chapter 14
I Make the First Great Breakthrough

Because I did not know what was about to occur, I was continuing on - contemplating the ramifications of particles really being holes in three dimensional space:

I was asking myself what seemed to be just a simple question, "...if three dimensional space is flowing into one hole and out of the other, which one does it flow into, and which one does it flow out of?

Without knowing what I was about to discover, I reasoned the solution to this problem might be found in the density of the space surrounding a hole. If three dimensional space were flowing out of a hole, it was only logical to conclude the density of the three dimensional space immediately surrounding the hole would be greater than the density of the three dimensional space further away. This effect would occur because the three dimensional space flowing out of a hole pushes against the three dimensional space already there, creating a spherical region of greater density, decreasing further away from the hole.

Figure 14.1

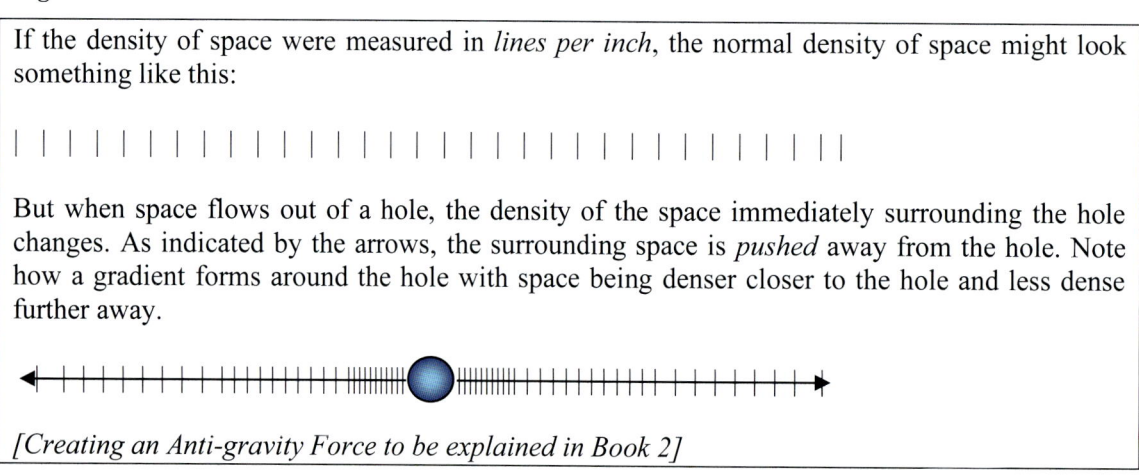

> If the density of space were measured in *lines per inch*, the normal density of space might look something like this:
>
> | | | | | | | | | | | | | | | | | | | | | | | | | | | | | | | | | | | | | | | | | | | | |
>
> But when space flows out of a hole, the density of the space immediately surrounding the hole changes. As indicated by the arrows, the surrounding space is *pushed* away from the hole. Note how a gradient forms around the hole with space being denser closer to the hole and less dense further away.
>
> *[Creating an Anti-gravity Force to be explained in Book 2]*

The opposite would be true for space flowing into a hole. If space were flowing into a hole, the space in the region directly surrounding it would be "pulled" - stretched into the hole. This pull would create a spherical region of less dense space immediately surrounding the hole; a region that slowly begins to return to its normal density much further away from the hole.

Figure 14.2

> Density of space surrounding a hole that *space flows into*:
> As indicated by the arrows, the surrounding space is *pulled* into this hole. Note too how the gradient formed here is less dense closer to the hole and denser further away.
>
> *[Creating a Nuclear Gravitational Force surrounding the proton: to be explained in Book 2]*

When I considered which bend would be the best candidate for creating the source of gravity, I realized it had to be the bend surrounded by the less dense region of space. Within this less dense region, space would be "stretched inward". This less dense region would make the space surrounding other holes stretch towards it: making the hole move in that direction too. Therefore, since I had previously concluded protons and neutrons were responsible for creating the less dense space surrounding stars, it was only logical to conclude three dimensional space was flowing into the proton, and out of the electron.

Even though this idea was contrary to the present scientific beliefs of how the lines of electrostatic forces are pointing, I already knew that the direction of the positive and negative electrostatic forces were originally and arbitrarily assigned by Benjamin Franklin. Nobody knows for sure just which direction is which. [It should also be noted, the mathematical proof of this theory, presented in the APPENDIX of the next book, will work no matter what direction space flows.]

"Wait a minute!"

Suddenly, I didn't care about "which direction was which." I didn't care about anything. Nothing mattered except for...Yes... There it was, right in front of me, the explanation for one of the greatest mysteries in all of science - the particle and wave characteristics of matter. It was unbelievable.

What was once perceived to be a very complicated idea was now rendered amazingly simple. The long trail of thought which led me to deduce the existence of less dense regions of space surrounding the proton and the electron now explained something totally unexpected.

Although many people nowadays understand that photons of energy possess both particle and wave characteristics, not many people are aware of the fact that matter does too. It seems to defy logic that matter should behave as both a particle and a wave, and yet it does. As strange as it seems, matter possesses characteristics which makes it behave both like a rock held in the hand, and sometimes like the waves it creates upon the surface of a pond after it is thrown in.

In the past there was no explanation for this dual phenomenon, but now there is: while the "holes" in three dimensional space are responsible for the particle effects of matter, the less dense (or denser) regions of space surrounding them are responsible for the "wave" effects of matter. Clearing up one of the great mysteries of science.

Figure 14.3

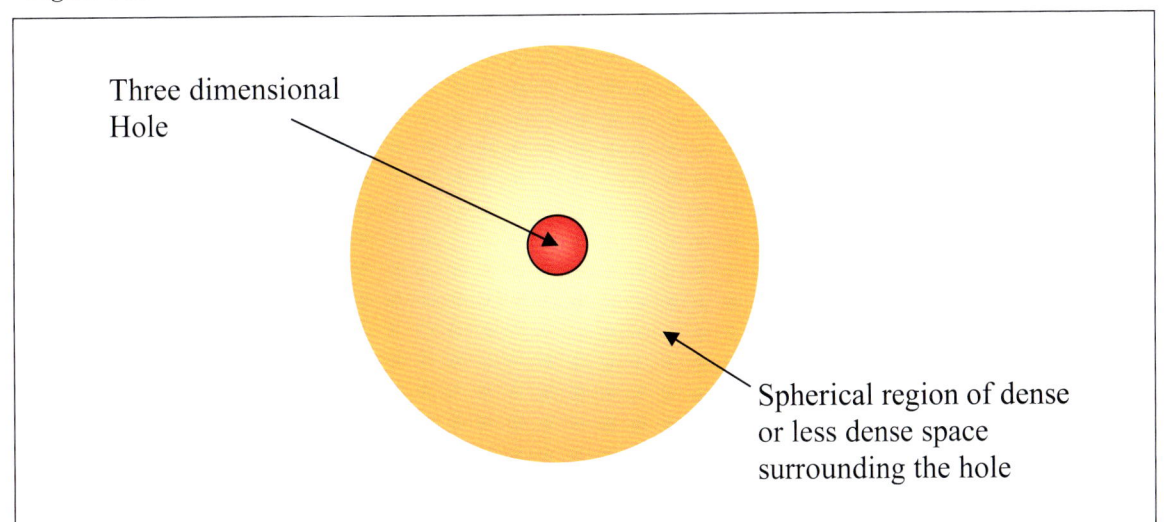

Three dimensional Hole

Spherical region of dense or less dense space surrounding the hole

Note: these proportions are not correct. The size of the spherical region of bent space surrounding individual particles appears to be immense. If the hole were the size of a bowling ball, the region of less dense space extends outward at least ten miles.

Chapter 15
I Discover How Atoms Are Really Constructed

I couldn't believe it! Through the conscious use of this new way of thought, intentionally, I had unintentionally stumbled upon the answer to one of the great mysteries of science.

The paradox of the particle and wave characteristics of matter has long been one of the most perplexing phenomena in the whole of physics. Many experiments have been performed to try to understand it; millions have been spent on research; and the scientific discussions on this one subject alone could fill many volumes. And yet, with this new way of thinking and reasoning, the explanation of the particle and wave characteristics of matter was simple, almost trivial.

This was exciting. Finally, at long last, I was on to something. Something big!

Many years before, I had reduced everything in the universe down to five separate parts: matter, space, time, energy, and the forces of nature. But now with a new way of thinking and reasoning, I had accomplished something that should be impossible. I had *united* two of these seemingly independent pieces - matter and space!

The implications of this discovery were shocking. Could all of the independent pieces of the universe actually be made out of only one thing, One Substance?

The only way to find out was to try to unite another piece. But what piece? What could I try next?

I decided to try energy. But while thinking about it, I unexpectedly made another great discovery.

I started by thinking about atomic explosions. I knew atomic explosions are supposed to be the proof that matter and energy are equivalent (energy can be created out of the destruction of matter). Consequently, if matter was created out of bent and flowing space, maybe energy was too? But how? How was a hole in space and energy related?

I set out to find the answer by first trying to envision what was happening to the three dimensional space flowing into the hole called the proton and out of the hole called the electron. What happens to this space? What happens to it after it flows into the proton, and before it flows out of the electron?

Although each question just seemed to raise another, the next question was actually the answer to all of them: since protons and electrons are in close proximity to each other in atoms, could the space flowing into one be flowing out of the other?

And then it came to me - vortices. Of course, miniature tornadoes of three dimensional space flowing from protons to electrons in higher dimensional space.

Figure 15.1

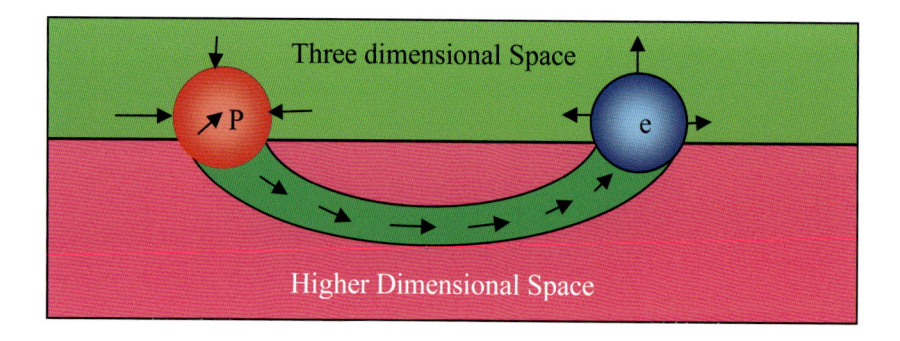

What a beautiful vision: an unimaginably small tornado of three dimensional space flowing between particles in higher dimensional space. [As will be seen in the second book of this six part series, this most important discovery explains many of the totally bizarre and unexpected phenomenon of the sub-atomic world.]

Suddenly I stopped thinking about energy; stunned, I realized this wonderful discovery suddenly explains the creation of atoms themselves!

Because three dimensional space is flowing into the proton and out of the electron, when these two "particles" come into close proximity, a unique situation develops.

Although present day science believes the electrostatic charges of protons and electrons pull these two particles together, this is an oversimplification. What really happens is that some of the space flowing out of the electron begins to flow into the proton. As these two particles move closer together, a critical distance is reached where *all* of the three dimensional space flowing out of the electron flows directly into the proton. When this situation occurs, *two* vortices of whirling space are created.

The first vortex is created in our three dimensional space. It flows from the electron to the proton. The second vortex is in higher dimensional space. This one flows from the proton to the electron in the opposite direction to the one flowing in three dimensional space.

These two vortices now create a circulating flow containing a fixed volume of space.

This circulating volume of three dimensional space flows from the proton, into fourth dimensional space - through fourth dimensional space, and then back into the electron. Here, it exits the electron, flowing back through three dimensional space and into the proton once again- binding the proton to the electron, creating a hydrogen atom.

Figure 15.2

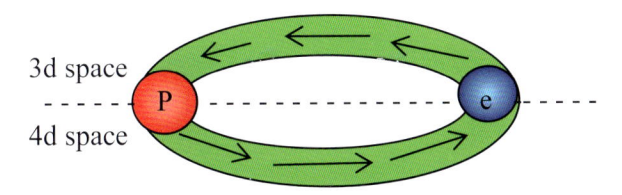

When the circulating flow commences, both of the electrostatic charges are neutralized. The word "neutralized" was used because no flowing space escapes from the system. If surrounding space still flowed into or out of this system, atoms would possess electrical charges (and every time we touched something we would get shocked).

Because these vortices exist in all atoms throughout the universe, the revolutionary new vision of matter presented within this book was christened *The Vortex Theory*.

This knowledge of the internal structure of the hydrogen atom has an added bonus - incredibly, it finally allows us to understand how a photon of light is constructed!

Chapter 16
Energy Is Finally Explained Too

We all know that everywhere we look in nature we see an incredible variety of colors and shapes. However, what most of us don't know is that we never see anything. We don't even see the words on this page.

The only things we ever "see" are "photons" of energy.

Photons of energy are emitted from an energy source (such as the Sun, or a light bulb), hit the page, "bounce off", and hit our eyes. Without these photons we wouldn't be able to see anything at all. But just what are we "seeing"? What are these "photons" of energy? (What is energy?)

One of the greatest mysteries in all of physics is energy: just what is it?

According to present day science, energy is contained within tiny particles called photons. These photons possess both particle and wave characteristics. But where they come from, what they are made of, or how they are constructed is unclear.

But not anymore.

With the discovery of the miniature vortices flowing back and forth between protons and electrons, the great mystery of the photon is finally unraveled - and a fascinating vision unfolds before us.

What contemporary science calls a photon is really a condensed portion of three dimensional space thrown out of the vortex.

Although there are several ways to create photons, in the hydrogen atom, a photon is part of the volume of three dimensional space flowing from the electron to the proton. When certain conditions occur, the vortex shortens, discharging a packet of condensed three dimensional space back into the three dimensional space from whence it came.

What is very important, is the photon's velocity. The velocity of the photon is the velocity of the vortex. Consequently, the speed of light can now be traced to the speed of the vortex.

Although the present vision of the universe presumes but cannot explain why the release of the photon from the atom somehow makes it "instantaneously" travel at the speed of light, we can now envision what is really happening. Using the principles of the vortex, it is easily understood that a photon is but a part of the volume of the three dimensional space already circulating at the speed of light before it was thrown free of the atom. It travels at the "speed of the light" because the space in the vortices is traveling at the "speed of light".

These principles can all be seen in the following drawing:

Figure 16.1

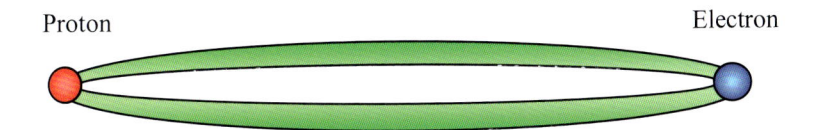

Proton — Electron

Vortex begins to shorten; photon begins to emerge from electron

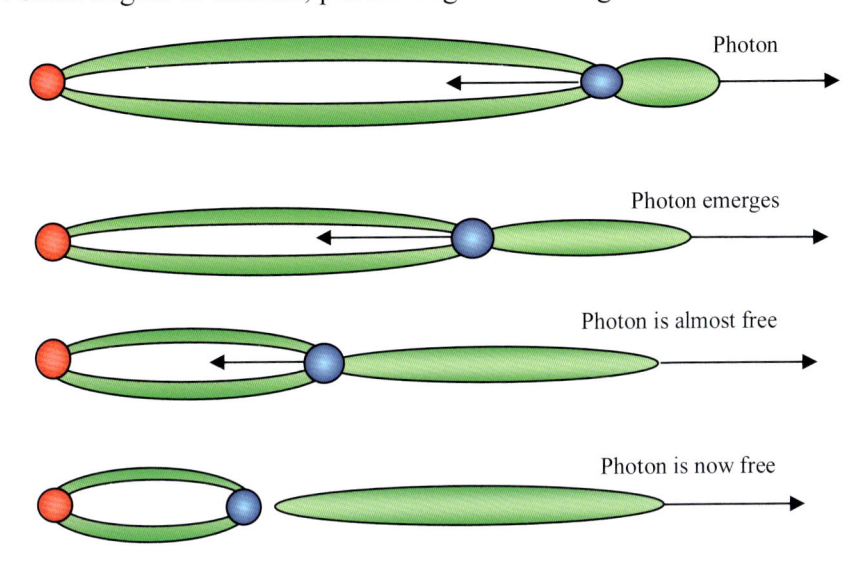

Photon

Photon emerges

Photon is almost free

Photon is now free

The photon can now be defined as a packet of condensed three dimensional space, expanding and contracting in a long tubular shape perpendicular to its direction of travel. The rate of its expansion and contraction – or frequency - is a direct function of the volume of condensed space within the photon. The particle effects of the photon are a direct result of its being a condensed packet of three dimensional space, while its wave effects are created by the expansion and then the contraction of the three dimensional space it is passing through. And just as before, using the principles of "less dense and flowing space", another one of science's greatest mysteries - the particle and wave effect of light - is easily explained.

(The reason why the photon expands and contracts will be explained in the second book of this six part series.)

If I was excited before, I was ecstatic now. I had united three parts of the universe that, up until now, were three separate and individual pieces. I was getting closer; I was closing in on the ultimate and final vision of the universe. However, I still had a big problem - the neutron. How was the neutron constructed?

Although the neutron should be discussed in the section under matter, the discovery of the vortex had to be made before the amazing secret of the neutron could be unraveled. The words "amazing secret" were carefully chosen because what comes next is unlike anything anyone has ever seen before.

Chapter 17
The Secret of the Neutron

The subatomic discoveries I had come across so far could be described as bizarre, fantastic, and unbelievable. But the next discovery tops even these descriptions.

It is easy to see how electrons and protons are three dimensional holes. This discovery is revealed because their electrical charges are created by three dimensional space flowing into or out of them. But the neutron has no charge. This means no space is flowing into it or out of it. So how can it be a hole?

The answer is that the neutron is not just one hole; it is a hole within a hole! A simply fantastic concept.

The neutron is created when an electron is shoved up against a proton and completely encircles it; or the vortex flowing out of a proton into higher dimensional space is hit by the right type of "particle", breaks and completely encircles its three dimensional surface. [This will be explained in great detail in Book 3 of this six part series.]

Because the electron completely encircles the proton, the space flowing out of the electron is no longer flowing outward into the three dimensional space of our universe. Instead, its direction is reversed. It is now flowing inwards, directly toward the three dimensional hole (the proton) the electron is surrounding. This situation creates an enclosed loop.

The space flows out of the proton and into higher dimensional space; as soon as it does, it fans outward into a cone shape, is turned inside out, and instantly curls back upon itself creating a tight loop. This tight loop completes the return back into three dimensional space by flowing directly onto the surface of the encircling electron, forming a fourth dimensional torus - or donut. A fantastic shape.

Perhaps even more fantastic than the shape of the neutron is its speed of circulation. This circulation is taking place within the shortest distance imaginable, yet moving at the incredible speed of 186,282 miles per second.

The neutron has no charge because none of the space surrounding the neutron flows into it, or out of it.

Also, the higher dimensional vortex is bent into a very tight loop. In this tight loop, the vortex is turned inside out, creating the weak force of nature [why the torus breaks – seen as the weak force - will be explained in Book 2]. Hence, the neutron's "neutral" charge, and the weak force of nature are both revealed, clearing up two more of the great mysteries of nature.

Because this extraordinary creation of nature we call the Neutron is so unique, I have attempted to illustrate it in the following drawings. Unfortunately, since it is impossible to draw fourth-dimensional space, these two dimensional to three dimensional sketches are used:

Figure 17.1 INITIAL CONDITION:

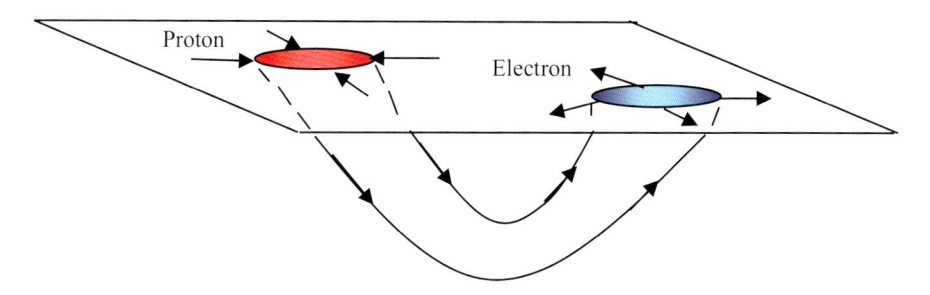

Figure 17.2 STEP #1: For any one of a number of possible reasons, the vortex breaks: [Note, this break is much closer to the surface than seen here.]

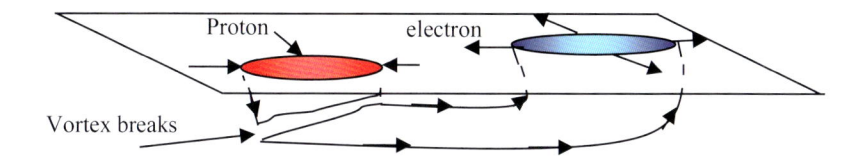

Figure 17.3 STEP #2: Isolating the proton from the above drawing, and expanding its size, note how the bottom of the vortex begins to curl outward:

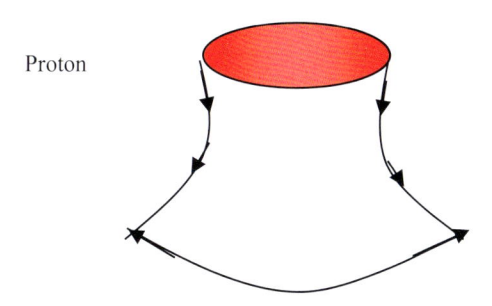

Figure 17.4 STEP #3: The curl becomes more pronounced as it continues to move upward:

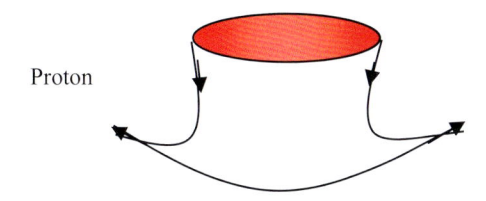

Figure 17.5 STEP #4: The vortex curls upward at an incredible speed (speed of light) towards the top of the hole we call the proton:

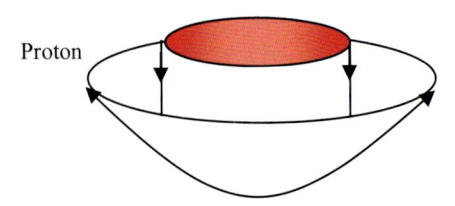

Figure 17.6 STEP #5: The vortex approaches the top of the hole called the proton:

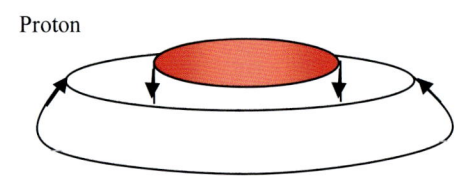

Figure 17.7 STEP #6: The vortex curls back into the hole called the Proton, forms a torus, the circulating flow begins and becomes a new "particle" science calls the Neutron.

Note: for scientists, in Book 3, the changes in the quark content from the proton to the neutron is explained.

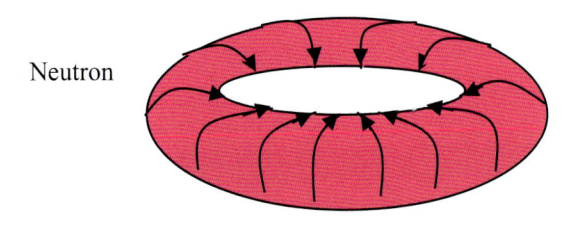

Figure 17.8 STEP #7: Another new "particle" called a *positron* is created when the end of the vortex attached to the electron reaches the two dimensional surface.

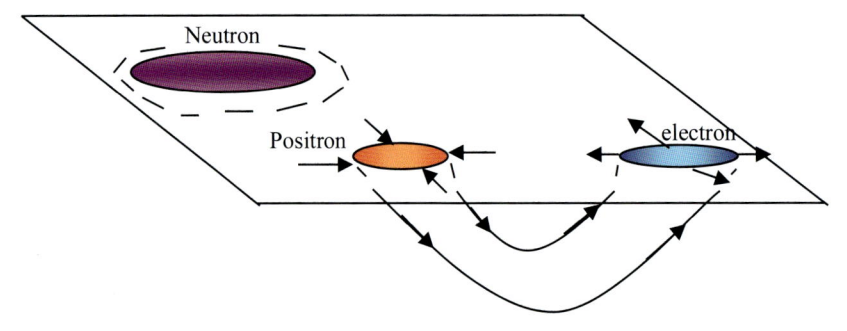

[Note how space now flows into the hole called the positron; turning it into a "particle" with a charge opposite to that of the electron. (For those with a physics background, *the free end of the vortex now explains the creation of the phenomenon called Conservation of Charge. And another of the great mysteries of science is no more.*)]

[Also, in Book 3, the explanation of how the quarks change "Flavor" from the proton to the neutron is explained.]

Chapter 18
The Four Forces of Nature Are Finally Explained

"Hallelujah, I was on a roll."

Like the mysterious pieces of the ultimate jigsaw puzzle, the four parts of the universe were becoming better defined, almost ready to reveal the ultimate vision of the universe. But it was a strange puzzle.

Instead of a jigsaw puzzle where you connect all of the pieces together, this was a puzzle where each piece somehow became something else. Where each time you connected two of them together, they disappeared into each other and became one - and you were always left with only one piece – space! It was a bizarre and perplexing experience. I didn't care.

It was the thrill of a lifetime. I was doing what no man had ever done before and would never do again. I was going to be the first person in all of human history to "see it all." And I was almost there.

I was very close to the "top of the mountain" where I could finally "look over" and see everything in the physical universe. But there was one more obstacle to overcome - force: the four forces of nature.

After eliminating time, I had managed to unite matter, space, and energy. The only piece of the universe left was force - the four forces of nature.

Could they be united too? Could this theory do what Einstein couldn't do?

Of the four forces, I realized I only had the strong force left to explain. Gravity was explained by the existence of less dense space between particles of matter. Einstein's hypothesis that bent space is equivalent to gravity is wrong. *It is less dense space that creates Gravity.* The bent space surrounding stars is itself a phenomenon - an effect - created by the spherical shell of less dense space surrounding large astronomical bodies.

[As mentioned before, in Book 2, both the Nuclear Force surrounding the proton, and the Anti-gravity Force surrounding the electron will be explained.]

The Electromagnetic force could be explained by flowing space; and the weak force could be explained by the breaking of an inverted tight loop of flowing space. So, it was a very good possibility the strong force could be explained by some sort of configuration of *dense and flowing space* too.

So, what was it?

Unfortunately, I had no idea.

For the first "time" in a long time, I was stumped. I had no answer and absolutely no idea of how to proceed.

At a loss to know what to do, I decided to review what was known about force.

I again read how the 20th Century scientific vision of the universe sees the four forces of nature to be made of something called "Force". But what force is, or how or why it even exists is still unknown. [Although the Theory of Quantum Mechanics claims forces are passed back and forth by particles, the Graviton, a hypothetical particle supposed to carry the force of Gravity has never been discovered.]

I read how the creation of the forces of nature has long been one of the great mysteries of the universe. When Isaac Newton discovered Gravity, he called it a "force". Since then, three more "forces" were discovered. And many physicists believe these four forces are all manifestations of just one universal force, but they can't prove it.

I also reviewed how many scientists including Albert Einstein tried to unite the four forces into one universal force. But Einstein, and everyone else who tried, failed miserably.

Of course, I knew now why they had failed. None of them knew "time" was a false concept, nor did they know about the principles of less dense and flowing space. If they did, perhaps things might be different.

But none of this research seemed to help.

After spending a lot of effort researching the past failures of others, I was still getting nowhere. So, I decided to try something else. I knew from studying quantum mechanics that the forces of nature are theorized to be the result of tiny particles. Thinking about the strong force and particles, I remembered something about a Japanese Physicist in the 1930s (Hideki Yukawa) proposing the strong force might be created by a "Virtual Particle" (an unseen particle being passed back and forth between the proton and the neutron).

However, from my knowledge of *dense and flowing space*, I knew such a particle would really be a three dimensional hole in space.

Of course. That was it!

The key phrase was, "…being passed back and forth…." Back and forth! In the blink of an eye, I solved the problem. And it was simple. It was so simple a child could understand it.

As you recall, the proton is a three dimensional hole surrounded by a spherical region of space bent into it, and even though the neutron is a combination of a hole within a hole, it is also surrounded by a spherical region of space bent into it.

When a neutron approaches a proton, these two depressions in space try to bend into each other. When they do, some of the space flowing out of the spherical hole surrounding the neutron tries to flow into the proton sitting beside it.

This situation would cause the enclosed vortex encircling the neutron to break free and encircle the opposite proton: turning *it* into the neutron while the one it vacated again becomes a proton. However, just as soon as the switch occurs, the process would begin all over again, causing it to reverse itself; the encircling vortex would break free of the proton it now surrounds, and return to the one it just left. It would keep on doing this dance, back and forth. Because this reversal happens so incredibly fast, this dancing hole of space would end up keeping one "particle" pressed tightly against the other - becoming the strong force of nature. [Note: it will be explained in Book 3 how UP & DOWN quarks shift back and forth between the two particles - creating the appearance of a "Virtual Particle": a Pion.]

Figure 18.1

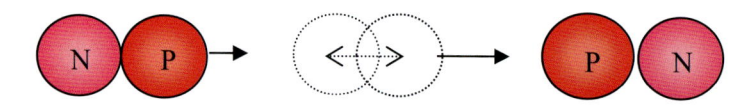

It should also be mentioned that since it is the invisible higher dimensional vortex that is first encircling one proton and then the other, from our perspective in three dimensional space, this would look like a three dimensional spherical particle is being passed back and forth.

This continual sharing or exchange of the three dimensional hole surrounding the neutron becomes the Strong force of nature, explaining another of the great mysteries of science.

[Note: sometimes, when two positively charged particles collide, the attempt by three dimensional space to form only one surface appears to be the reason why charged particles can temporarily bond together forming the Delta particle - a <u>very</u> <u>short</u> lived particle with a charge double that of a proton. This will be covered in Book 3 of this six part series.]

And suddenly there it was.

I had used the principles of *dense and flowing space* to explain the four forces of nature. And most wonderful of all, using the principles of *dense and flowing space*, the four forces of nature could now be united.

It was another great day in my life. A day I dreamed of and lived for. Everything in the universe, including the forces of nature could now be explained by the principles of *dense and flowing space*. Hallelujah, what a discovery. I was ecstatic.

But this was just a prelude. An anti-climax. Because the greatest discovery was yet to come, and it was even more spectacular because it was unforeseen - totally unexpected.

Chapter 19
The Shocking Truth About Force!

Ever since Isaac Newton got hit on the head by an apple falling from a tree and discovered gravity, the world has had to deal with the presence of a mysterious unseen "force". This force called Gravity causes matter to accelerate towards other matter. Even when we are standing still upon the seemingly motionless surface of this planet, the molecules within our bodies are constantly being accelerated towards it. We call this acceleration our weight. It is this acceleration that holds us upon the surface of this spinning sphere and keeps us from flying off into space.

Because of the above facts, we live with force every day. Force is with us every second of our worldly lives.

Because three more forces were eventually discovered, the word "force" has become a casual part of our vocabulary. We use this word freely and easily even though we don't realize the creation of force is still a total and complete mystery. [Even the highly accepted and widely believed particle explanations of force proposed in Quantum Mechanics is a mistake as we shall soon see.] Although we all believe in force, nobody [until now] knows how it is being generated.

Although at one time, nobody in the entire world knew about the force of gravity, today, there is perhaps nobody in the world who does not know about it. It is another one of those very few commonly shared beliefs accepted among all the people of this planet.

I know about it because nobody believed in gravity more fervently than I. Before I undertook this journey of discovery, I believed force was a mysterious energy or substance surrounding matter, binding it together.

Because of this misconception, it took a while before I finally realized the shocking truth about force. Just like the "shocking truth" about matter, space, time, and energy is like nothing we have ever imagined before, so is force:

The shocking truth about "force" is that it just doesn't exist.

The reason why nobody can explain what force is or how it is being generated comes from the fact that the universe contains no mysterious "substance" called force. Nor are there any subatomic particles that "carry force" with them as proposed in the Theory of Quantum Mechanics.

[NOTE (for scientists or scientific minds): the photon doesn't "carry" the electromagnetic force, it merely adds to and extends the vortices; the weak force is not transferred by the "W" particle. Instead, the "W" particle is nothing more than a vortex untwisting back into its normal shape after being "abnormally" twisted into a 4d torus (like that in the neutron); nor does the "Gluon" transfer the strong force. Instead, as we have seen above, it is the continual reconfiguration of the fourth dimensional torus moving back and forth between the proton and the neutron that is holding them together; and finally, as it will be shown below, the force of gravity is not being created by the graviton. Instead, it is being created by the distortion of the shape of a hole as shown in Figure 19.1]

If the great scientists of the 20th Century (including Einstein) who tried to unite the forces were able to see exactly how the universe is constructed; they would be astonished to learn that there are no "forces" of nature. They were trying to unite something that doesn't exist!

What we call gravitational force (also in the electromagnetic attraction and repulsive "force") is merely a condition created by dense and flowing space causing the normal spherical shape of particles to distort into "pear shapes". These pear shapes are unnatural bends. This unnatural bend creates a surface tension that tries to straighten itself back into a sphere. As it does, the sphere (hole) moves slightly in the direction of the other hole. However, as soon as it starts to straighten out, it is again distorted, repeating the process, making it move faster and faster, accelerating the particle towards the other hole. [Note, in the case of the Electrostatic Repulsive Force, the direction of the distortion is reversed, and the particles move away from each other.]

When we stand upon this planet, the shape of every proton, electron, and neutron is distorted towards the Earth's center. The collective attempt of these particles to straighten themselves out, *pushes us* towards the surface of the Earth {*we are not attracted to the earth as is presently believed*!} and becomes the "force" we identify as our weight.

Figure 19.1

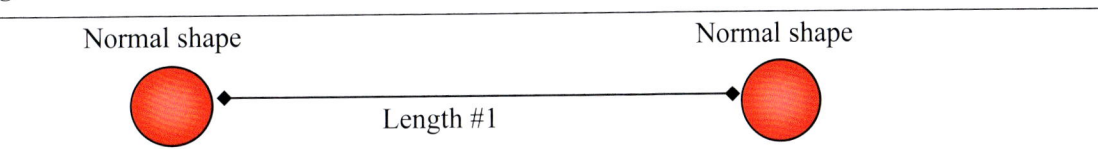

Normal shape   Normal shape

Length #1

When close together or in a massive region of less dense space (powerful gravitational field), the spherical shape distorts into a pear shape:

Distorted shape   Distorted shape

At the point of the pear, the hole is sharply bent. As it tries to straighten back into a spherical shape, the backside of each hole moves towards the point. This motion shifts each sphere towards the other, accelerating them towards each other.

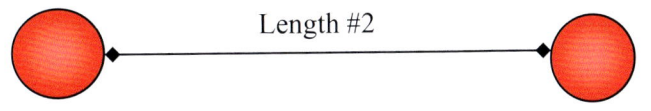

Length #2

However, as soon as they become spheres, they distort, and the process happens all over again:

As the process of distortion and reshaping repeats itself, note how length #2 becomes shorter than length #1. This change in distance represents the motion of these two holes as they move towards each other. The density of space determines how fast they accelerate towards each other.

Note: although this entire process will be explained in greater depth in Book 2 of this six part series, a little thought will enable anyone to figure out how this same process works with the other three "forces". Also, a little more thought will explain why a hole surrounded with space *bent outward* is *also* accelerated towards a region of less dense space.

The knowledge of the creation of the phenomenon of force is the fourth and final piece of the puzzle. It is the union of these four pieces that completes the picture and finally allows us to see the ultimate vision of the universe:

When I first saw the true vision of the universe, the vision no man has ever seen before, I was awestruck. Thunderstruck!

This true vision of the universe is the greatest discovery ever made. It is a shocker. A blockbuster:

Stated simply, there are no separate "parts" to the universe.

There is only space and its motion. [And that's it!]

The only thing that exists in the physical universe is the substance of which space is made. Space is not "space" at all. Space is made of something totally unique from our point of view. It is constructed out of at least seven dimensions, and it can both bend and flow. [The why and the necessity of seven dimensions are discussed in Book 3]

This substance is in motion. It is expanding outward at an incredible speed carrying the Galaxies with it. Both matter and energy are created out of space. The "particles" of matter - protons, electrons, and neutrons - [and a host of other "particles"] are not particles at all. They are three dimensional holes existing upon the surface of fourth dimensional space. The substance of which space is made flows into and out of these holes.

The hole creates a particle effect, while the denser or less dense space surrounding the hole creates a wave effect.

Energy is also created out of dense and flowing space. Energy in the form of photons, are really denser regions of space that expand and contract as they move. This dense region - or photon - creates a particle effect, while the expansion and contraction of the space it passes through creates a wave effect.

Time does not exist; the phenomenon of time is created throughout the universe by the uniform flow of microscopic space into and out of the three dimensional holes of matter.

There are <u>NO</u> "forces" of nature. Using the massive expansion of three dimensional space throughout the universe as a reference, its backward flow into the tiny, microscopic holes is responsible for the creation of four special, though uniform configurations of dense and flowing space. These uniform configurations distort the spherical surfaces of the three dimensional holes, accelerating them towards each other, creating the illusion of force.

Chapter 20
At Last - The Alpha and the Omega

When seen for the first time, the vision of our incredibly diverse and complex universe being created out of only one substance is a shocker. And yet strangely, for the first time in many years, it also left me feeling calm and at peace with myself.

At long last, the awful feeling of not knowing the answers and the helplessness of not being able to find the answers was gone. I had finally freed myself from the curse of ignorance that possesses the entire human race.

I had crawled out of the abyss. I had gone where no man has ever gone before, and I had seen the vision of the universe that no man has ever seen before.

But what had I seen? And even more important, what does it mean?

Like Columbus of old, had I stumbled upon a brand new world while trying to find the old East Indies. A brand new world indeed!

The above discovery is so revolutionary there is nothing to compare it to. It stands alone and apart from all of the other knowledge in the world. *It is the beginning and the end, the Alpha and the Omega of discoveries.*

Everything that exists, every physical relationship, even the creation of the universe itself can be explained by this vision. {The creation of the physical universe will be explained in Book 2 of this six part series.} Hence, the ultimate vision of the universe is also - *"The Theory of Everything" The Rosetta stone of science.*

This vision also possesses a delicate elegance unique unto itself. Its wonderful simplicity surrounds it with an aura of beauty, unlike anything ever seen before.

To grasp the incredible implications of this vision, first look at or visualize all of the different colors you can see or think of. Look at the reds, the blues, the yellows, the greens, and all of the myriad numbers of colors in-between – and then realize they are all made out of just one "color" possessing different vibratory rates. [All photons are probably a silvery color.]

Next, visually imagine anything and everything you can think of. Think of the forests, the mountains, the deserts, the oceans, the white clouds in the deep blue sky, or the twinkling stars in the dark night sky and realize this incredible diversity of shapes, sizes and textures are all made out of just one substance. It is a humbling and profound experience.

But there is even a more profound experience: look around the room you are sitting in. Look at all of the different and separate things you see including *yourself*. Now look again at these separate objects and realize *these are not separate objects at all*.

EVERYTHING YOU SEE, INCLUDING YOURSELF, IS ONE THING. THERE ARE NO SEPARATE PARTS.
SEPARATION IS AN ILLUSION. IT IS AN ILLUSION CREATED BY THE FACT THAT WE CANNOT SEE THE MATERIAL SPACE IS MADE OF. WE ONLY SEE THE HOLES AND NOT THE MATERIAL BETWEEN THE HOLES.
EVERYTHING THAT EXISTS EVERYWHERE WITHIN THE PHYSICAL UNIVERSE, INCLUDING EVERY ONE OF US, IS MADE OF ONE SUBSTANCE.
WE ARE ALL ONE WITH IT. WE ALL MOVE WITHIN IT. WE ALL EXIST WITHIN IT.

Wow!

I wish I could adequately express all of my feelings when I discovered this ultimate vision of the universe, but any effort seems inadequate. Someone once said, "that man's creations pale beside the works of God". And no truer statement was ever made.

This vision of the universe is also peaceful. The amazing complexity of the physical universe being created by something so incredibly simple is comforting and reassuring. In fact, the more you think about it the calmer you will become. Try it yourself.

The next time you are stuck in rush hour traffic on the freeway, imagine the peacefulness and the stillness of the substance of which everything you see around you is imbedded within. Imagine its quietness, its calmness, even as the chaos of matter seethes *within it.*

When I first did it, it was as if I was at one with the universe; it was meditation; it was the OM: it was a religious experience. A most profound religious experience.

A most profound religious experience! Wait a minute. A most profound religious experience?

What does this mean? What was I really looking at? What had I found? Is it God?

Chapter 21
Is This God? Have We Discovered God?

A great philosopher once said, "Know thyself".

I guess I never knew myself until I asked myself the question, "Is this God?" because when I did, I was suddenly afraid.

Just what had I stumbled onto?

I was just a mental adventurer in search of the thoughts no man had ever thought before. But never, never in my wildest dreams, did I ever even think I might discover God.

Had I?

How can such a question even be answered? How *would* anyone know if they had discovered God? How *could* anyone know if they discovered God?

Is there any way to recognize God? How can God be identified? What is the definition of God?

Looking in the Bible for a description of God we see terms such as the "*Alpha and the Omega*", and the "*I am, That I am*".

Using these terms, it is easy to see that the substance out of which space is made is certainly the Alpha and the Omega, the beginning and the end of everything that exists in the physical universe, but is it the "I am, that I am" of the Bible?

Just what is it? Is there a way to find out?

Grateful for the opportunity to turn my mind to another subject, I looked for a way to resolve this dilemma. And while thinking about it, I realized the answer could be found in the creation of the universe.

I realized that if God created the universe, there are three ways he could have done it: the universe is created out of God; the universe is a unique manifestation of God; or the universe itself is a creation of God.

If the universe is created out of God, the space of the universe is God, and we exist within the actual physical body of God. This vision is comparable to the existence of a single biological cell living within our physical body. (A single microscopic creature existing within a much larger, massive creature.)

If the universe is a unique manifestation of God, the space of the universe might be a thought and we might be mental projections within this thought form. Although this idea might seem farfetched, we must never forget that we dream. In our dreams we appear to see real images, and have real experiences. [How does this happen, how are these images formed, and why do they appear so real?]

But if the universe itself is a creation of God, then space is a creation of God, and we are creations within the creation. This scenario can be compared to a man inventing a liquid plastic, forming bubbles within its volume, and then somehow arranging the bubbles to form images.

So, which is which?

Another question: can a mere mortal man even answer such a question? Are there limits to the mental constructs our mind is capable of forming?

The mental picture of God being indescribable and incomprehensible to the mind of a mere man is one of the accepted beliefs of world religion. But is it true? Can a creation analyze its creator? Was it a sin to even try?

But then again, maybe it would be a sin to not try. Maybe I was reacting to the inherited fears of our ancient ancestors. The same fears that kept them from crossing the ocean because they were all afraid they were going to fall off of the edge of the world. Or the fears that kept them from challenging the hypocritical doctrine of the church leaders, and allowing the inquisition to take place because they believed these maniacal priests were doing the will of God.

In the past, men also believed in demons. They tortured and murdered other men in fear of them. However, they failed to realize the only real demons were the fear and ignorance within their own minds.

Fear and ignorance, ignorance and fear. The true rulers of men. Allowed to run rampant and without restraint they accentuate each other, intensify each other, until they dominate the minds of men.

But no matter how powerful they seem; they are only thoughts. Because they are only thoughts, the way to beat them is with another thought or with no thought at all: to totally ignore their presence.

And that is exactly what I did. I told myself that my fears were mental projections put into my mind by ignorant men and gave them no more credence. Then, I boldly returned to the subject of the moment: is the space of the universe God?

While thinking about it, I began to recall all of the religious texts of the world. I remembered that wherever men believed in God there seemed to be three words everyone everywhere in the world agreed upon. Three words universally accepted and associated with God in all religious beliefs. These words are *omnipresent, omniscient, and omnipotent*. They speak for themselves:

If the substance out of which space is made is God, then God is surely *omnipresent*. God is everywhere. Every "particle" of matter and every photon of energy everywhere in the physical universe is made out of him. Since he would be aware of every motion of everything that exists everywhere in the universe, he would know "when a single sparrow falls from the sky". He would also know the configurations and the interconnections of every neuron within a man's mind. He would know when electrical current flows between them, and in doing so, he would know every thought a man thinks.

Knowing every thought a man thinks, there would be no fooling God. He would know the hypocrisy, the lies, and the true feelings of every man in the world. Knowing all and seeing all, he would know the secrets of every person everywhere in the universe.

He would know when a man earnestly prayed to be heard by him, or to really be seen by other men. He would know what happens in the light of day or in the darkest dungeon in the world. His knowledge would be all encompassing, and universal. He would indeed be *omniscient*.

However, is he *omnipotent?*

It is easy to see how the substance out of which space is made is omnipresent, and omniscient, but to be omnipotent is another matter entirely.

The deep contemplation of this problem reveals that the key to its answer is found in motion. The creation of the universe began when space was put into motion. But did space set *itself* into motion, or did something else do it?

If something else set it into motion, it is not omnipotent and it is not God. However, if the substance out of which space is made set itself into motion, then *it* created the universe out of itself. It made everything, and in doing so, became omnipotent.

To be omnipotent, means it had the ability to create the universe and used that ability. This shows premeditation and cognizance of thought. Which means space had to have the ability to both think and reason before the physical universe was created.

However, all of these ideas are all predicated upon the premise that space possesses awareness - consciousness. Does it?

While thinking of a way to answer this question, surprisingly, the following answer seemed to just pop into my mind all by itself. It begins like this:

Where does our awareness - our consciousness - come from? In fact, where does any animal's awareness come from?

If we could take a human being apart, atom by atom, we would find we would have several thousand piles of compounds. If we could then break these compounds apart and separate them into the individual elements they are constructed out of, we would have about a hundred piles of individual elements. And finally, if we could take these individual elements apart, we would end up with just three big piles of electrons, protons, and neutrons.

But these are inanimate objects. They possess no consciousness or awareness. Furthermore, all three of these piles are not "matter" at all. They are just three massive collections of three dimensional holes in space. So, we are left with a stunning conclusion: a particular arrangement of three dimensional holes creates consciousness!

And something even more fantastic: the inanimate objects of the universe have obtained consciousness of themselves. When *we* look at a proton or an electron, in reality, it is the matter of the universe that is looking back at itself. A simply astounding idea!

And it is this astounding idea that leads us to another even more astonishing idea: if a creation made out of nothing but holes in space can obtain consciousness of itself, can the space it is created out of obtain consciousness of itself too?

If it can, and if it set itself into motion creating the physical universe - and us in the process - then my friends, it is God.

PART III

Chapter 22
The Search for a Proof Begins

"Wow again!" In a previous chapter I used the word "Wow" (an expression of amazement). But even a hundred "Wows" seems inadequate when talking about possibly discovering God!

There are a lot of mental discoveries that ignite and boost our excitement. But the possibility of finding God had the opposite effect upon me. I became wary of myself. Had I gone a "bridge too far" in this mental journey? Was I mad?

I needed to get back to reality (whatever that is). Although the question, "Space possesses consciousness?" is not rhetorical, it does not appear to be provable or "disprovable" either. I know because I tried to think of every way possible to do it. Afterwards, I regrettably knew it was time to return to the beginning. Why the beginning?

After realizing I had finally come to the end of a simply wonderful chain of revolutionary discoveries, I knew I needed to return my attention to what started it all - higher dimensional space. Because of all the shocking discoveries I had made, perhaps the most shocking of all was the fact that I couldn't believe any of it - yet.

I had accumulated a lot of circumstantial evidence for the existence of higher dimensional space and its implications. Metaphorically speaking, I had gone where no man has gone before. I had stood upon the mountain and had seen the wonders of the universe the world's greatest philosophers, scientists, and wise men have only dreamt of seeing.

I felt privileged and humbled. But, it was all for nothing unless I could prove it.

Although I desperately wanted to believe in higher dimensions where the Kingdom of God existed, where souls existed as souls, and where peace, joy, love, and happiness reigns supreme forever and ever, I also knew I couldn't believe any of it unless I had proof. Just like the Disciple Thomas who needed to see the holes in Jesus' hands before he could believe, I needed proof too. And to convince me that Heaven existed, I was going to have to see one Hell of a proof!

Here is where I became really tough upon myself. Because not only did this proof have to reveal the Vortex Theory was true - it also had to prove the Vortex Theory was equal to – and equivalent to the Theory of Relativity. It had to show the Vortex Theory explained the bizarre phenomenon of the universe just as well as the Theory of Relativity.

This meant I was going to have to somehow use the Vortex Theory to explain the "granddaddy" of all the special effects in the universe – the strange and curious length shrinkage, and time dilation effects that occur at near light velocities.

To accomplish this seemingly impossible feat, I was going to have to come up with a mathematical explanation for the way atoms shrank and the phenomenon of time slowed down when traveling at near light velocities. To do this, I was first going to have to discover the mechanics of how and why the vortices flowing between the protons and electrons shrank when atoms begin to move.

Simply put, I had to do what no man had ever done before, I had to successfully find another explanation for the most difficult problem man has ever encountered in his exploration of the universe. I also knew there would only be me to do it. I would not get any help from anybody else.

Faced with the enormity of such a problem, I was temporarily overwhelmed.

My god what a problem! Why did it have to fall to me? Who was I? I was not an engineer or a mathematician. I was a construction worker now. I built pools for a living. I didn't solve complicated math problems anymore.

Math problems! What an understatement! This was no ordinary math problem. The true explanation of the Michelson Morley Experiment was a world class math problem. It was the Olympic Games of math problems. People won Nobel prizes for solving math problems such as this one. Men who are now legends in the world of science were the ones who originally made the previous explanations of this experiment; and there were also many other legends in the world of science who tried, but were incapable of solving this problem.

How could I hope to compete with this crowd?

But then I had another thought.

Although at first glance, such a task might seem impossible, I knew if the Vortex Theory was the correct explanation for the construction of the universe, the proof already existed; it was already there, just waiting to be discovered.

So, in the midst of despair, I realized I had nothing to fear. If the Vortex Theory was correct, it would explain exactly how the length of objects shrank, and how the phenomenon of "time" slowed down when traveling at near light velocities.

I had traveled a long way using that little quote from Jesus in the New Testament. It would be a pity to quit now.

Since I had found the rest, I made up my mind I would find the proof too; I made up my mind I was going to succeed; that I had already succeeded. Besides, if I was right, how could I fail? So, get going Russell. What's your problem? Hang in there buddy. Don't quit. Never give up. Never. Never. Never!

(Reluctantly, I set out to find the answer.)

Little did I know it would take seven more years.

Seven more difficult and frustrating years.

Many times I tried and failed. But I never quit; I never lost faith I would succeed. I kept on and on, until eventually, I triumphed. And it was a wonderful success.

Using that little quote made by Jesus in the New Testament, I discovered one of the most beautiful and elegant mathematical relationships in all of nature. I discovered mechanical movements of the vortices taking place within atoms that are so precise and so exact, there is nothing in our contemporary world to compare them to. And it was in these mathematical relationships and mechanical movements of the vortices where I discovered exactly how matter shrinks, and how the motions of matter slow down - making it appear as if "time" itself has somehow slowed down.

These wonderful explanations finally allowed me to deduce the true explanation for the Michelson Morley Experiment. The true explanation of this most important of all scientific experiments proves the existence of higher dimensional space – which proves that two thousand years ago Jesus knew more about the construction of the universe than all of the greatest scientists who have ever lived!

This mathematical proof can be found at the end of this book.

It should also be mentioned this proof was just the beginning of many more wonderful discoveries. These discoveries were so numerous they take up five more books.

The one discovery that needs to be mentioned here is so important it cannot wait. This discovery explains one of the greatest secrets of the universe - *the secret of time* - the secret of how "time itself" is created (This secret is explained at the end of this section.)

Chapter 23
What Does It All Mean?

The end of the era of time is the end of the longest and most enduring era in all of human history.

Although "time" was just an idea, the end of time is not just the end of an idea; it is the end of a way of life. Over the countless millennia that time has been a part of the collective consciousness of man, the entire culture of the human race has become tied to this concept, and everything that can be thought of is affected by its demise.

The end of the era of time affects all of science, technology, philosophy, and religion. Because these four subjects encompass every field of human endeavor, the lives of every person in every culture and in every country in the world will be directly affected.

As previously mentioned, every science book in the world is now obsolete. Equally shocking is the fact that every school, college, and university is teaching all of their students scientific knowledge that is totally and completely wrong.

All of the science teachers, engineers, and technicians in the world need to be reeducated and retrained. But when they are, they will not only encounter the revolutionary knowledge that will allow them to see the ultimate vision of the universe, they will also encounter the following dilemma:

> "In an age when there were no scientists or scientific experiments, how did Jesus possess the scientific knowledge of the universe which no man in the world knew until now? Not just any knowledge, but knowledge of the actual existence of higher dimensional space. Knowledge which is perhaps the most difficult of all the scientific knowledge of the universe to acquire, because it can only be proven <u>after</u> the acquisition of a long chain of other scientific discoveries.
>
> *Also, it is very important to know that the existence of higher dimensional space is not some sort of low priority subject; without the presence of higher dimensional space, NOTHING within the three dimensional universe would exist. There would be no three dimensional vortices extending into higher dimensional space. Because there would be no three dimensional holes there would be no protons, electrons, or atoms. Consequently, higher dimensional space is not some insignificant place existing in a little corner of the cosmos. It is the most important place in the entire universe. Without higher dimensional space there is no universe, no stars, no Earth, and no us.*
>
> Furthermore, the discovery of the existence of higher dimensional space and the remarkable chain of scientific discoveries that followed; all resulted from the words of Jesus in the NEW TESTAMENT. Without the words of Jesus, these discoveries might not have ever been made."

There is no denying the above fact or escaping from it. One quote, spoken 2000 years ago by a supposedly "disgruntled Jewish philosopher", has suddenly turned the logic of science upon itself. Just like a sun-burnt scorpion stings itself to death, the very principles of science, which once turned men against the belief in God, will now turn men against the "scientific atheism" these principles were responsible for creating.

The circle is now complete – science has discovered the "Backdoor" to the Kingdom of God.

Although it will take some time for this astounding conclusion to become part of the collective consciousness of man, when it finally does, something amazing will have occurred. A scientific revolution will have swept through the churches of the world, and a religious revolution will have swept through every scientific center of learning in the world.

Each will complement the other, and together, the whole will become greater than the sum of its parts. And just as the intellectuals of the past rejected Jesus, the intellectuals of the future will accept Jesus. Perhaps they will make the world a better place. But one thing is for sure, they will be faced with one of the greatest philosophical problems ever to confront mankind: do we exist within the body of God?

PART IV

The proof of this revolutionary new vision of the universe is found in its precise mathematical explanation for the length shrinkage and time dilation effects associated with the Michelson Morley Experiment. Named the Vortex Theory, this new vision of the universe proves why the length of matter parallel to the direction of travel shrinks; then it proves why matter in the transverse direction to the velocity of travel does not shrink. And finally, it gives an exact mathematical explanation for the phenomenon of time dilation. This proof can be found at the end of this book. This proof was also published in 2012 by the Russian Academy of Sciences: it is number [1] listed below.

Also, two experiments were performed to physically prove the theory is true: they were published below in papers [2] & [3]:

[1] See number 8 on page 117
[2] See number 9 on page 117
[3] See number 8 in References on page 113

Chapter 24
The Secret of Time Is Finally Explained

The human race has lived with the concept of time for at least five thousand years. But until now, nobody has known what it is or how it is created. However, that has all changed.

Here, at long last, is the explanation of how "time" – or rather the phenomenon of time is created:

Although "time" does not exist as a fundamental principle of the universe, the phenomenon of time is created by a hierarchy of speeds beginning with the velocity of the space flowing into and out of the three dimensional holes of "matter".

Because it is easier to see this relationship when envisioning the two vortices of the three dimensional space flowing back and forth between the proton and the electron in the hydrogen atom, this example is used: [see Figure 15.2]

Earlier, it was shown how the speed of a photon of light is a function of the speed of the vortices. However, even this velocity is a function of something else. The velocity of the space flowing within the vortices appears to be a result of the fastest speed higher dimensional space is capable of bending and flexing.

This speed appears to be the fastest motion in the universe. Why the vortex possesses this speed will be examined in greater depth in Book 2 of this six part series. For now, it can be said this speed can be identified as the speed of light.

The speed of light is at the top or peak of a pyramid of speeds. Although the speed of light is the reference to which all other speeds are referred, the speed of light itself is a function of the velocity of the space flowing within the vortices.

The speed of the electron as it moves about the photon is also a function of the speed of the vortices. Since the electron is a hole created within three dimensional space by the flow of one of the vortices, its position is determined by the changing position of the vortex. Hence its speed as it moves about the proton is a percentage of the speed of the vortices.

Since the proton is also a hole in space which three dimensional space is flowing into, it cannot move faster than the speed of the space flowing into it. Consequently, its speed is also a function of the speed of the space in the vortices. The same is true for the three dimensional space circulating within the neutron.

Since all atoms are made up of protons, electrons, and neutrons, the speed of the motions of all atoms is again a percentage of the speed of the vortices. The same holds true for molecules.

Since all molecules are made up of atoms, the speeds and the motions, and the vibrations of all molecules are percentages of the speed of the atoms out of which they are made. Hence, a molecule's speed is also a function of the speed of the vortices as well.

And last but not least, at the very end of the hierarchy of motion, is the motion of planets, stars, and Galaxies. Because all planets, stars, and galaxies are made out of atoms and molecules, their much slower motions are also functions of the velocity of the space flowing within the vortices. (Note: the above is also true for antimatter atoms and molecules made up of anti-protons, positrons, and anti-neutrons.) This hierarchy of motions now allows the phenomenon of time to be explained.

Since all motions associated with the passage of time (including those created within clocks) are created either by light, or by the motions of matter, it is now easy to see that all of these motions are also functions of, and percentages of the speed of the space flowing within the vortices. Because all the vortices of all the atoms in the universe are flowing at the same measured rate, the same motions within every atom of every similar element and molecule throughout the universe also take place at the same measured rate.

It should be noted here that if the vortices were flowing at different speeds at different places in the universe, the speed of the photons of light thrown from these vortices would be variables instead of constants. This would make the light coming from distant suns and galaxies travel at different velocities. Since this is not seen, the velocities of all the vortices are the same throughout the universe. [And as will be seen in the third book of this six part series, this deduction gives us a clue as to how the universe was created.]

These similar motions are responsible for the creation of similar harmonic motions occurring everywhere at the same rate. These similar atomic, chemical, biological, and astronomical motions, no matter where they occur in the universe, create an orderly sequence of constantly repeating events. These constantly repeating events create order and harmony in the universe. **This order and harmony is responsible for the creation of the phenomenon of time.**

It can now be understood that all similar motions occur everywhere at the same rate not because of some metaphysical quality called "time", but simply because the *dense and flowing space* out of which everything in the universe is made moves at the same rate everywhere.

Because it could also be misconstrued that time does indeed exist and is in fact controlling how fast space can bend and flow, this mistaken idea must be dispelled. Realizing the "bend-ability" of space is a function of its elastic properties dispels this idea. This "bend-ability" is an important subject because it relates to the "solid-ness" of space.

The topic of the "solid-ness" of space came up when I was once told by a professor of physics… "space could not be made of anything because it would have to be extremely dense to allow light to move through it so fast". Consequently, if it was very dense it could not flow – making all of the ideas about *dense and flowing space* false. However, this type of reasoning is incorrect.

The speed of light is a function of the velocity of the vortex. Consequently, because we measure the speed of light with harmonic motions that are also functions of the speed of the vortex – there is <u>no</u> motion in the universe that is independent of the speed of the vortex. Therefore, any attempt to measure the speed of the vortex has to be made with another motion that, unfortunately, is also a function of the speed of the same vortex we are trying to measure. Hence, if the vortex moves slower, the velocity of the motion used to measure it also moves slower.

This deduction leads us to another astounding conclusion: the speed of the vortex [and photons of light] could really be one foot per second, or one inch per second, or one millionth of an inch per second and we would never know it!

Because of this amazing fact, the length of a second – which seems very short to us, might really be very long. In fact, every motion in the universe might be so slow it could have taken me a year to write this one sentence. Then again, a second might really be incredibly short. It is possible it could have only taken a microsecond to write this entire book. Either way, we would never know it. We will never know it because no matter how fast or slow the vortices are moving, it will always appear to us and our instruments that they are moving at 186,000 miles per second – the speed of light. A simply astounding conclusion!

Chapter 25
Time Dilation Is Finally Explained

Perhaps the most fascinating phenomenon existing in the universe today is "Time Dilation".

When an object begins to move faster, from the 20th Century scientific view, its relationship with "time" somehow slows down. This is a real effect possessing astonishing results. The most famous of these effects is the "Twins Paradox":

Say for example when two twin boys were about eighteen years old, one of them left the earth for a long voyage on a spaceship to a distant star. This spaceship, traveling at .866C, which is nearly the speed of light (note, .25C is one-fourth the speed of light, .5C is one-half the speed of light, and .75C is three-fourths the speed of light. "C" stands for light speed.), reached the distant star and immediately turned around and came home.

Because the entire round trip took forty years, when the spaceship returned to earth, the brother who was now fifty-eight years old was there to meet him. However, when the spacecraft door opened, instead of him greeting another fifty-eight year old man, he was amazed to find his brother had only aged twenty years.

Even though the earthly calendar indicated he was gone for forty years, the calendar on the ship indicated he had been away for only twenty years. Amazed, they both realized somehow, some kind of "time dilation" effect had occurred. Yet neither could explain how it happened.

Neither could anyone else in the world explain how it happened - that is until now. Here is how it happened.

When the spaceship left the earth, and begins to travel at a near light velocity, the atoms the ship is made of were moving at a much SLOWER velocity than the atoms the earth is made of.

This means the electrons, protons, and neutrons of every atom within the ship are moving slower too. And again, just like before, *even though the vortices connecting every proton and electron continue to appear to flow at precisely the speed of light, from the point of view within a proton, electron or neutron,* the speeds of the vortices appear to change. One vortex appears to be going much faster while the other appears to be going much slower.

The addition of these two speeds creates a slower round trip time for the space circulating within the vortices. Making the round trip take longer! Making the same round trip take place at what can be called, "A SLOWER APPARENT VELOCITY." [Note: the same effect occurs for the space circulating within the neutron.]

Also, because the spaceship is moving, space is constantly reconfiguring itself around every tiny three dimensional hole the ship is made of. As each individual hole moves, the space surrounding it is first bent towards it, and then away from it as it passes by. So even though a proton, electron, or neutron may not be attached to an atom, the space surrounding it ends up moving in two opposite directions, making it reconfigure around the particle at the rate of the slower apparent velocity too. Furthermore, because everything in the universe is constructed out of *dense and flowing space*, including energy, and the "forces of nature", all of their motions, and all of their interrelated motions, will also slow. In fact, the mathematics reveal that all of their motions will slow down exactly and proportionally to the correspondingly slower apparent velocity of space when they are traveling in the faster moving frame of reference.

For simplicity's sake, the spacecraft was said to be moving (in the mathematics of the thesis) at a velocity of .866 the speed of light. This speed was chosen because at this velocity, the round-trip time of the vortices of every atom within the ship doubles. And when the round trip time doubles, and the reconfiguration of *dense and flowing space* doubles, the "distance" between every motion of every atom within the ship also doubles. Because of the hierarchy of motions described before, when the motions of atoms double, the motions of the objects made out of them double too. Making every motion now takes *twice as long* to complete.

Consequently, when the ship was sitting upon the earth, if it took one second to throw a baseball from one man to another within the ship, when moving at .866C it now takes twice as long. Hence, the ball itself now moves at a slower velocity that is directly proportional to the SLOWER APPARENT VELOCITY of the atoms it is constructed out of.

Also, since the atoms in the man's arm are now moving at the slower apparent velocity of space, it cannot move as fast as it could when the ship was motionless. Therefore, the man cannot throw the ball as fast as he could when the spaceship was not moving. However, since the motions of every atom in the ship have slowed down proportionally to the slower apparent velocity of space, NOTHING appears to have changed. If the man was capable of throwing the ball at a speed of one hundred miles an hour when the ship was motionless, to all of those within the ship, the ball still appears to be moving at a speed of one hundred miles an hour. From the perspective of all those within the ship, everything "is as it always was". But this is just an illusion.

In this faster moving frame of reference, the motions of everything have now slowed down proportionally to the motions of everything else. This means EVERY motion, including all motions within all assorted types and kinds of clocks, now takes place at a slower rate than it did when it was at rest in reference to the earth.

Hence, a clock attached to a wall in the interior of the spaceship now runs slower in reference to a clock on the earth. It runs slower because every motion of every atom within the clock now takes twice as long as it did before. When it used to tick off two seconds, it now ticks off only one second. One hour becomes a half-hour, and 24 hours turns into 12.

[If an atomic clock is launched into space aboard this spaceship, orbits the earth at a high rate of speed, and is then brought back down to the surface; comparing its time to the time on another atomic clock will make it seem as if this clock has run slower. (Which it has.) However, "time" was not the cause of this "Time Dilation" effect. Instead, the culprit was the slower apparent velocity of *dense and flowing space*. Allowing us to see the real cause of this time dilation experiment mistakenly used as a former proof of Einstein's Theory of Relativity.]

Returning our attention to the spacecraft, because all motions have slowed down proportionally to all other motions, the biological processes within the traveler's body have slowed down too. This means that in addition to the motions of the atoms and molecules within the traveler's body, his very thought processes will also have slowed down proportionally to the slower motions of the atoms within in his body. Since the electric current and chemical reactions within his body have slowed down, his thoughts and perceptions have slowed down, making it seem to him as if nothing within the ship is any different from what it was before.

However, everything *is* different. Motions have not only slowed, but the shapes of everything within the ship are now radically distorted, but amazingly, everything still appears to be normal.

Everything still appears normal because of the greatest optical illusion ever created. If we could see past the illusion, we would view what can only be called *THE REAL AND FRIGHTENING VISION OF THE UNIVERSE.* A vision where everyone would appear to shrink in the direction of travel, making them all look like card-board cut-outs! (This startling vision will be revealed in Book 2 of this six part series.)

Chapter 26
The End of the Theory of Relativity

The era of the Theory of Relativity comes to a close when it is realized Albert Einstein's vision of the universe is completely wrong. It is wrong because the Theory of Relativity is based upon the effects of the universe and not the causes.

There are two completely different visions of the universe: the vision of effect, and the vision of cause. The first vision is seen with our eyes, while the second vision is perceived with our minds. And even though each appears real, the first is nothing but an illusion created by the second.

The idea that space is made of nothing, matter is made of something, and time exists as a fourth dimension is totally wrong. Nothing could be further from the truth.

Even though Mr. Einstein's fourth dimension of "space-time" was a brilliant idea, his use of it to explain the length shrinkage and time dilation effects associated with near light velocities was dead wrong. Einstein was close to the truth, but his use of the fourth dimension was more of a metaphysical explanation than a precise mechanical explanation.

Although there is a fourth dimension, it doesn't possess any "time" characteristics. It doesn't somehow create length shrinkage and time dilation effects. Instead, when matter moves, it is the shrinkage of the vortices themselves flowing back and forth between three dimensional space and higher dimensional space, which are creating the length shrinkage effects. And as can be seen in THE PROOF (Book 2), it is the slower apparent velocity of *dense and flowing space* that is creating the "time dilation" effects.

Einstein once said, "Imagination is greater than education". A profound insight which is absolutely true. For when everyone is trained alike, everyone thinks alike, everyone reasons alike, and everyone comes to the same conclusions.

This continued way of thought does not find errors, but reaffirms errors from one generation to the next. The only way to break this misleading cycle is to throw away the authoritarian educational training received from others and to use one's own imagination to think in ways no one has ever thought in before.

This was the very way of thinking that originally launched Einstein upon a voyage of discovery, but never allowed him to reach the "New World": (never allowed him to see the true vision of the universe). Mr. Einstein's problem was that he tried to incorporate the concept of time into all of his ideas, never realizing that this was the excess baggage of past generations which needed to be discarded before ever trying to make the trip.

Although he was right many times, he was also wrong.

Einstein's use of his imagination allowed him to correctly deduce the speed of light would be measured the same for all observers everywhere no matter how fast they were traveling. A brilliant deduction. However, because Einstein believed space was made of nothing, he never discovered or deduced the principles of the vortices. Hence, from Einstein's point of view, there was no standard to measure anything to. Consequently, the motions of everything in the universe appeared "relative" to the motions of everything else. A situation that appears to be absolutely true for the observations of the planets, stars, and galaxies we see from our perspective as human beings living here upon the Earth. Hence, the Lorentz transformation equations that allow us to calculate the "time differences" between two moving frames of reference are still valid. And even though the "twin paradox", the orbit of Mercury aberrations, and many other observations from the "relativistic" vision of the universe are still real effects, the causes of these effects have nothing to do with the concept of time or of a fourth dimension called "space-time".

Just as Newton's laws are still applicable even though they were amended by the Theory of Relativity, the Theory of Relativity is still applicable too, even though it is now amended by the Vortex Theory.

However, in the "microscopic" world of the atom, things are very different. The "microscopic" universe possesses a completely different perspective than that seen from the "macroscopic" point of view of the universe.

The view of the universe from inside a proton or an electron is completely different from the relativistic view. Because of these two differences in perspective, there will always be two truths to our viewpoint of the universe. There is the truth of cause and the truth of effect.

The misunderstanding of these two truths has fooled many of the greatest thinkers who have ever lived. Aristotle was fooled. Aristotle believed everything above the orbit of the Moon was eternal and unchanging. An observation which from his perspective [and from ours], appears to be absolutely correct. When we look up at the positions of the stars in the sky, they do indeed appear to be forever eternal and unchanging. A mistaken belief that makes his assumption appear to be absolutely true. Indeed, it is a true effect - we actually see it. In fact, from the time we are born until the time we die, nothing appears to have changed in the night sky. We can truthfully witness and testify to this effect in court.

Even so, nothing could be further from the truth. Everything in the sky is constantly changing its position. What we are witnessing and testifying to are only the effects we are seeing from our point of view here upon the earth. Today, we all know the cause of this effect results from the fact these stars are so incredibly far away their actual motions are indiscernible.

Another example of these two contrary truths is seen in the rotation of the heavens. From the viewpoint of every generation up to and including Copernicus, it appeared as if everything was revolving around the Earth. Although the stars do not appear to move as individuals, "en-mass" they appeared to revolve as one group - along with the Sun, Moon, and planets around the Earth. Which is a true observation! However, as before, this is only the effect we are seeing. As we now know, the cause of the effect is being created by the rotation of the Earth itself.

Cause and effect - it crops up again and again in man's observations of the universe. We see an effect and try to assign a cause to it, but many times the causes we assign are completely wrong. But this technicality doesn't seem to bother most people. People need answers, and if one fulfills this need, whether it is right or wrong, the uncertainty is eliminated, and men are satisfied.

When their children ask them the same questions they asked their parents, they are happy to be able to give them an answer. Soon the answer becomes part of the collective heritage of the ideas passed on from generation to generation. It becomes a belief and becomes a part of our mythology. The more people who believe it, the more believable it becomes.

In knowing how errors are made and propagated from generation to generation, we now understand Mr. Einstein's error. Like everyone else, he was taught to believe in the erroneous concept of time. Since he believed in time, he believed the cause of the time dilation effects occurring at near light velocities was "time". However, after the discovery of the Vortex Theory, we can now see that there is a difference between the "effect" of time and the "cause" of time.

Although the "effect" of time - [the phenomenon of time] - appears to be real, it is not being created by "time". Instead, the motions of everything within the physical universe are creating it. However, it must be realized these motions are themselves a function of the ELASTICITY OF

SPACE. A characteristic of space causing it to move at the same rate everywhere throughout the universe.

This similar rate creates a "synchronous repetition" for harmonic motions. Making all similar motions of matter occur at the same rate everywhere. Therefore, anything which might change the elasticity of space, will also change the rate at which motions occur, creating the effect of "time dilation". Hence, the intense "gravitational fields" created by the massive regions of less dense space existing around stars, changes the elasticity of space - making it move slower - creating time dilation effects. Allowing us to finally understand why "gravity" affects time. Answering the last of the enigmas of Relativity.

Consequently, in lieu of the above knowledge, the Theory of Relativity is now but another chapter in the history of mankind. An era gone by. An era noted for its nuclear bombs and atomic powered submarines. Another era attributed to the worst and the best which mankind has to offer. An arrogant era where physicists told the people of the world Albert Einstein's Theory of Relativity was absolutely right, and anybody who disagreed was either uneducated or a fool. Well, at least they didn't burn people at the stake like their predecessors did.

What a time.

Although it is sad to see the end of an era, it is exciting to contemplate the beginning of a new one. An era which promises to be mankind's greatest. Because with anti-gravity engineering, we will be able to get off of the earth and go to the stars.

EPILOGUE

Many years had passed since I read the story in *Readers Digest*. I had come a long way. I had discovered the ultimate knowledge of the universe; and I had found a way to prove it to even my skeptical mind. But when I tried to give it to the world, I ran into the ultimate "Catch 22".

Because I refused to edit the words *God* and *Jesus* from the math, no physicist would even look at it. Because I could not get the backing of "physicists", no science journal would publish this work. [And even if they would, none of them would allow any religious connotations to be a part of an article in one of their journals.]

Because I could not get a science journal to publish my work, no newspaper or science oriented magazine would do an article on it. Without a panel of physicists to tell them the math was right; they were afraid of making fools of themselves by telling the world a construction worker found a mistake in Einstein's Theory of Relativity.

Television and radio were out. I contacted stations, and took my work to them, but they refused for the same reasons. And I would not degrade the words of Jesus by going on one of the daytime talk shows, so what to do?

Thinking about the problem and knowing there must be other people like me in the world who desperately wanted to know the answers but had no way to find them, I finally decided to write a short book (it was too long for an article).

However, thinking about writing a book, and actually writing it, are two different matters.

Because I had trouble just writing letters, it was a long drawn out process. After months of working on it at night and on weekends, I finished a primitive version, made ten copies and gave it to ten friends to read.

Because they were friends, they all told me, "Y'all, it's great. It's terrific. I really enjoyed it, y'all, really." But the look on their faces told the true story; none of them could tell me what happened past page three!

It wasn't very good.

Troubled and disappointed I said, "To hell with it! To hell with it all! Who even cares about the truth? If men choose to be dumb, let them be dumb. I'm finished with the foolish!"

So, I sat the book aside and forgot about it.

But I didn't know that other forces were still at work.

About six months later, on a Sunday afternoon, I was driving east on Sheridan Street in Fort Lauderdale, Florida. I was in my old pickup truck, heading back to my apartment on Hollywood beach, when a wheel bearing went out on the right front wheel.

Because all of the garages are closed on Sunday, I knew I could not get it repaired until Monday. Also, if I had it towed to where I lived, it would be very difficult or impossible to work on it in my narrow parking space. So, my next thought was to think about where I could keep it overnight.

Everyone I knew lived on the other side of town, except for one friend "Dee" who lived about two miles away, and she was one of the people I had given the copy of my short book to.

I had known her for many years. Her late husband and I used to be partners, and I knew she wouldn't mind helping out. I called her up and she said, "Of course it was all right". So, I called a tow truck. This "person" charged me fifty dollars to make the ten minute trip and I was boiling mad.

But Dee told me she was glad I wasn't in a wreck and to, "forget about the truck driver."

She then changed the subject by saying some friends of hers from England, who were staying there for a few days and were out at the moment would be back soon. So, I got out my tools, jacked up the truck, and started taking off the front wheel, when a jeep pulled into the driveway.

A man and a woman got out. I recognized the man as "Pablo", but the woman I had never seen before. Although she was plainly dressed, there was something very attractive about her. She had a warm and friendly smile and appeared to be about my age. I was told her name was Jeannie.

Later, I went into the back yard where Dee, Jeannie and Pablo were seated at a table, and sat down. After some small conversation had taken place between the four of us for a few minutes, Pablo and Dee got up from the table to move some roofing material stored on her porch leaving me and Jeannie alone.

Making conservation, I asked Jeannic what she thought of America. She told me it was a wonderful country. She explained how Pablo and his wife Janna were good friends of hers. About how she wanted to attend a healing conference in Tucson Arizona, but didn't want to go alone. That she enjoyed being in America, thought it was a great country and was sad she was going home tomorrow. And then she said something odd:

She said she was sorry she was leaving because she had read a wonderful book about "Time" just before she left England, and would have liked to meet the author before she went home.

She began to describe it, and as she did, it sounded exactly like mine. I couldn't believe it.

Surprising her, (and myself), I suddenly got up from the table, went out to my truck and came back with a copy of the book and put it down on the table.

"Was this the book?"

"Yes, that's it. That's the one I read."

"Well, I'm the author!"

We just sat there staring at each other.

I would like to have said something to her, but I was confused. Just how did she get a copy of my book and how did it get to England?

But I soon forgot about the book.

The more I looked at Jeannie, and the more she looked at me, it was as if we were old friends.

When Dee and Pablo came back to the table, and they both heard the story, Pablo confessed he was the one who had taken a copy of the book to England.

Six months before, when he was visiting Dee, he had seen the copy of the book on her desk. He had looked at it, found it interesting, and since he couldn't buy one, had secretly made a Xerox copy of it. He had taken it to England and shown it to his wife and some friends, including Jeannie. But she was the only one who had actually read it.

We were all amazed.

Jeannie asked me when I was going to have it published and I told her I was fed up with the book. I told her how everyone who was capable of understanding the mathematical proof was ignoring it because I was a construction worker who was inspired by the words of Jesus in the New Testament.

I told her it would probably never be published.

Later in the evening we had a chance to talk again in private.

She insisted it was too important a work to let die. She said I had to get it out to the people of the world because they needed it to overcome the atheistic attitudes dominating present day science. God needs to be put back into the scientific equations explaining the workings of the universe. That the very people who think they don't need it, are the very people who need it the most, and just like the "me" of many years ago, they need this knowledge to regain their faith in God too.

Ten months later, we were married.

Our friends said it was meant to happen. Everyone who has heard the story is amazed at the incredible synchronicity of the events occurring on that day.

They all keep marveling about it. What if I hadn't been driving on that particular road; or what if the truck had broken down in any other place; or what if she had not been there on that one day, we would never have met. Even more incredible, she wanted to meet me, and somehow, I miraculously appeared.

But I have another idea: it was another one of those events necessary to make the people of the world aware of the knowledge within this book. Without her, this book would never have been written.

She urged me on. She is the one responsible for having this book written as it is now and published. I would never have done it if it hadn't been for her.

As I look back at my life, it is almost as if I was an invisible pawn being pushed secretly across some massive chessboard in a mighty chess game between science and religion. An invisible pawn suddenly appearing miraculously at the end of the game, surprising the opposition, and checkmating the dark king of atheistic science.

And when I say I was pushed, I mean pushed. When I tried to quit, I was not allowed to quit. Every possible obstacle I encountered was overcome by me or for me. And when I was down and out in the last round of the fight, and thought I was finished, someone always miraculously appeared, gave me a hand up, and urged me onward. Ever onward. It is almost as if God wants the people of the world to know these ultimate and final truths about the universe and won't settle for anything less.

[I would tell you some more of these seeming miraculous events, including how I was awarded a PhD in Nuclear Physics by the Russian Ministry of Education… "But that is another story".]

The End

APPENDIX

THE PROOF:

MODIFIED PHD THESIS:

The knowledge presented in these 6 books is based upon the mathematics presented in this PhD Thesis. Although slightly modified to make it more readable for individuals without a mathematical background, this paper was awarded a PhD [called *A Candidate of Science and Engineering* in Russia] in Nuclear Physics in 2005 by the Russian Ministry of Education.

This thesis is being presented here to reveal the foundation for all of these scientific discoveries and to negate the arguments of so-called "debunkers", who over the years, have attacked this theory on the internet for a number of odd and bizarre reasons. However, even though they might not like the idea that space is made of something, or that Albert Einstein's theory of Relativity is made obsolete by this discovery, none of them has ever said anything about the mathematics.

This mathematics is confirmed and endorsed by none other than the powerful Russian Ministry of Education that award all PhD's in Russia. In 2012, this thesis was published by the Russian Academy of Sciences, peer review journal of branch in St. Petersburg State University:

Dr. Russell Moon, *The End of "Time"*, Book of Academic Papers, International Conference; Natural and Anthropogenic Aerosols VIII, October 1-5, 2012, The Learned Works Addendum, Part 2, 2013 Saint-Petersburg Russia, 2013; pages 473-488; VVM Publishing House 2019; 534 pages ISBN 978-5-9651-0804-6

The Thesis of the Mathematical Proof is Presented
The End of "Time"!

Dr. Russell G. Moon PhD, Nuclear Physics
Independent Researcher, USA
December 31, 2018

Abstract

Using the principles of the Vortex Theory, a revolutionary explanation for the length shrinkage and time dilation effects associated with the Michelson Morley experiment is presented. It is discovered that the phenomenon we call "time" is created by two miniature vortices of space flowing back and forth between protons and electrons in atoms. Although there is a fourth dimension, it possesses no time characteristics. Length shrinkage and time dilation effects occurring at near light velocities are caused by changes in the lengths and apparent velocities of the vortices. The slower apparent velocity of space reconfiguring around particles of matter [such as muons] as they move at near light velocities results in delay of decay exactly in correspondence to the Lorentz Factor. The less dense space surrounding planets and stars is not only responsible for gravity, but for slowing the motion of atoms and photons of light creating the time dilation effects associated with gravity. Leading to the revelation that time is only a function of motion, a phenomenon created by motion and does not exist upon its own as a fundamental principle of the universe.

Keywords: Time; Time Dilation; Vortex Theory; Michelson Morley Experiment; Relativity; Einstein; Gravity

1. A Curious Philosophical Question

If every motion of everything in the universe stopped and then started up again, is there any way to tell for how long it was stopped? After much contemplation, the answer is no. Every natural or manmade way used to keep track of time is in motion, [even tree rings are created by the movement of particles in chemical react-ions]. Although this is merely a philosophical question, for science, it possesses profound implications. For thousands of years man has believed in the concept of time; but if time stops when motion stops, time may not be real.

Instead, what we call "time" might really be just a function of motion, a phenomenon created by motion. But even more important, if time is not real, Einstein's 4th dimension of "space-time" does not exist. (A 4th dimension of space appears to exist.) But if "space-time" does not exist, something else is creating the time dilation and length shrinkage phenomenon associated with near light velocities. Consequently, there has to be a third explanation for the Michelson Morley Experiment: [the first by Lorentz and FitzGerald's who believed time was linear; and the second by Albert Einstein who believed time was relative].

This third explanation has to be totally unique: only the motions of the atoms themselves as they hurl through space can be causing length shrinkage and time dilation.

Does such an explanation exist? The answer is yes.

2. A Revolutionary Answer

The first clue to discovering the answer to our question was found in the construction of space. If "space-time" doesn't exist, the 20th Century vision of the construction of space is a mistake. Furthermore, since the construction of space also defines how matter is constructed, 20th Century Science's explanation of the construction of particles could be a mistake too.

If this is true, a revolutionary vision of space and matter must exist that can explain length shrinkage and time dilation without the use of "time": this led to the discovery of what has come to be called *"The Vortex Theory of Atomic Particles"*.

Using the Vortex Theory, a third explanation of the Michelson Morley Experiment was discovered. This explanation is presented in this paper.

[Because the Vortex Theory is relatively unknown, a brief synopsis of its explanation of how matter and space are created is presented:]

3. Background Information

SPACE: according to the Vortex Theory of Atomic Particles, space is made of something. This is *not* a return to the Aether Theory where it was mistakenly believed space was made of something and matter was made out of condensations of space: like ice in water. Because as matter moves, such a scenario creates an Aether wind: and the Aether wind was eliminated by the Michelson Morley Experiment.

MATTER: according to the Vortex Theory, "particles" of matter such as protons and electrons are not particles at all. Instead, they are hypothesized to be three dimensional (3d) holes existing upon the surface of fourth dimensional (4d) space. As these holes move through 3d space, 3d space reconfigures around them and no Aether wind is created.

The proton is a 3d hole bent into the surface of 4d space; 3d space flows into the proton creating its electrostatic charge. The electron is a 3d hole bent out of 4d space; 3d space flows out of the electron creating its electrostatic charge.

Figure 1

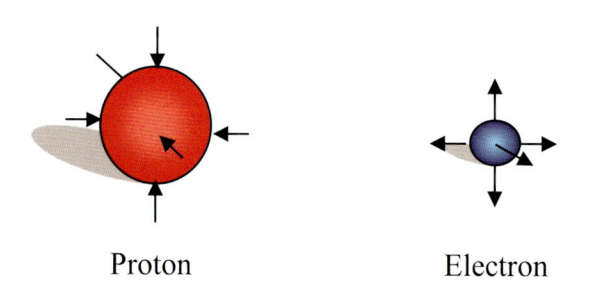

Proton   Electron

Because space is pulled into the proton it is surrounded by a region of less dense space. Likewise, because space flows out of the electron, it is surrounded by a region of denser space. These important concepts will be elaborated upon later.

[Note: although the quarks within protons are also explained by the Vortex Theory, it was subsequently discovered that the 4d vortex does not flow through them; hence, they are not necessary for this mathematical proof.]

Furthermore, it is hypothesized that these holes are really the ends of invisible 4d vortices existing in 4d space, connecting the proton to an anti-proton, and the electron to a positron. Three dimensional space flows into the proton, through a 4d vortex and exits out of the anti-proton; likewise, 3d space flows into the positron, through a 4d vortex and out of the electron.

The proton and the electron are connected by an invisible vortex of 3d space flowing from the proton to the electron in 4d space.

Figure 2

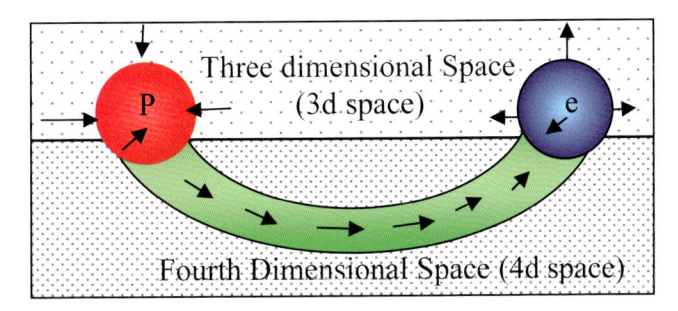

4. When a Hydrogen Atom is Created

<u>THE TWO VORTICES</u>: When a proton captures an electron creating a hydrogen atom, the vortices that connect them to the anti-proton and the positron break, reconnecting the proton to the electron, and the anti-proton to the positron.

If the proton and the electron are pulled closer to each other by their electrostatic forces, a second vortex is created in 3d space when all of the space flowing out of the electron begins to flow into the proton: creating a situation seen in Figure 3. Space now flows into the proton, into 4d space through the 4d vortex, back into the electron; then out of the electron and through 3d space and back into the proton: creating a circulating flow containing a fixed volume of space.

Figure 3

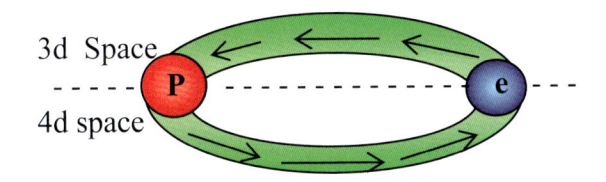

<u>THE HYDROGEN ATOM</u>: Because the hydrogen atom is the simplest of all atoms, it is the example used in this analysis. However, since all protons and electrons *in* all atoms everywhere would also be connected by two vortices of flowing space, the principles introduced here apply to all other atoms in the universe as well.

When the circulating flow commences, both electrostatic charges are neutralized. The word "neutralized" was used because no flowing space escapes from the system.

[Note: ions are created when two atoms in a molecule are separated and a proton in one atom is connected to an electron in another atom via a 4d vortex that still flows between them.]

5. Important Concepts

<u>THE ENCLOSED LOOP</u>: when a proton captures an electron, the electromagnetic force pulls these two holes together creating a hydrogen atom and the two vortices: 3d (three-dimensional) space flows from the electron through 3d space and into the proton; it then flows out of the proton and into 4d (fourth dimensional) space through the 4d vortex, and then back into the electron completing the circuit. This circulation creates an ENCLOSED LOOP. The circulating flow begins, and the volume of flowing 3d space within the two vortices remains trapped, creating an unchanging CONSTANT VOLUME.

This constant volume does not change. If it did, space would constantly be added or subtracted from the vortices causing all atoms to possess electrostatic charges. Hence, the volume of 3d space within this enclosed system remains the same no matter how fast the atom moves. In addition, because the radius of the atom is the length of one of the vortices, the radius of the atom now becomes a function of the maximum volume of space per unit of *measured* time that can flow from the electron to the proton or from the proton to the electron without backing up around either particle.

If space did back up, the speed of the vortices would drop below the speed of light and the electrostatic charges on protons and electrons would be variables instead of constants.

Finally, although the electron and proton are of different sizes, this in no way affects our calculations. Because the charges on electrons and protons are of the same magnitude, the same volume of space flowing out of one particle flows into the other. [In *The Vortex Theory*, it is shown how the space surrounding the proton is less dense. This region of less dense space allows its bigger, though less dense surface to have the same volume of space flowing into it, as that of the smaller diameter though denser space surrounding the hole we call the electron.]

6. The Formulas Used in this Proof

Note: because the proton and the electron both possess the same electrostatic charge, *the same volume of 3d space is flowing into and out of each hole*. Therefore, even though one particle is larger than the other particle, the cross sectional areas of the space flowing into and out of each particle are <u>*mathematically*</u> equal.

6.1. Calculating the Flow

Because the velocity into either particle equals the velocity out, the cross sectional area at any place on the length of either vortex *contains the same volume* of 3d space flowing through it. This allows us to calculate the volume of the flowing space that we call the electrostatic charge into and out of each particle in the following way: since the flow of a fluid is equal to the cross-sectional area multiplied times its velocity, then:

(1) $\qquad e = AV$

Where: e = flowing space = electrostatic charge; A = Cross-sectional area; V = Velocity

6.2. Units Used

Because the exact size of the electron is unknown, we do not know its exact surface area. Therefore, we cannot calculate the exact size of the vortices. Consequently, we can only calculate <u>the percentage the radius of the atom shrinks.</u>

To calculate percentage of shrinkage, actual units become irrelevant. Therefore, although the actual radius of the hydrogen atom is 5.29×10^{-11}m, it is assigned a value of 1m. Furthermore, because this paper theorizes that the electrical charges are actually *volumes of space* entering the proton and exiting the electron, their charges of 1.602×10^{-19} Coulombs are not rated according to cubic meters of space flowing per second of time and become irrelevant to our proof. Therefore, the cross sectional area of the vortices is assigned a value of $1m^2$ and the velocity of the vortex [the speed of light (C)] is reduced to 1m/s. This set of units equate to a volume of $1m^3$/s entering the proton and exiting the electron.

(Because we are only interested in how the atom shrinks at a velocity approaching the speed of light, the path of the electron about the proton is also irrelevant to our discussion. Only the position of the electron on the shell is important for our proof and not its momentum. How the electron reached that position does not affect our results nor violate the Heisenberg Uncertainty principle.)

6.3. Volume In Equals Volume Out

Because electrons and protons are similar to valves, the volume of space flowing INTO or OUT of either particle is found using formula **(1)** and is equal to:

(2) $\qquad\qquad\qquad V_i = V_o$

Where: V_i = volume flowing in = AV
V_o = volume flowing out = AV

For the purposes of our proof, the volume of the flow of 3d space into or out of either particle is equated to the constant value of $1m^3$. Unlike traditional valves, the flow of space into and out of electrons and protons experience no losses. If they did experience losses, the volume of space flowing between electrons and protons would constantly drop in magnitude causing the electron and proton to move closer and closer to each other until they eventually touched.

6.4. Factors Affecting the Volume of the Flow

Because the two vortices have created an entrapped circulating flow possessing a CONSTANT VOLUME, changes in their lengths create changes in the cross sectional area of the flow, changing the *measured volume of the flow*. [If the length is shortened, the cross-sectional area of a vortex increases, increasing the measured volume of its flow.] Therefore, to obtain the correct flow into or out of the electron or proton per unit of measured time the volume must be divided by the LENGTH of the vortices.

6.4.1. Dividing by Length and Unit of Length

The size of the atom is so incredibly small that changes in its length are meaningless to us unless we have something to reference them to. Therefore, by using a percentage of length change, we do not have to know the actual microscopic value of the length change that occurs. Instead, we can state that the diameter of the atom shrank by 50%, 60%, etc. Consequently, the length of the flow is divided by its unit of length to change it into a PERCENTAGE of the radius:

$$(3) \quad e = F_v / [L / L_U]$$

[Where: L=length; L_U=unit of length; F_v=flowing volume]

6.4.2. Multiplying by the Phenomenon of Time

Note: even though the explanation of "time dilation" will be discussed later, its effects are intertwined with length shrinkage and must be used here. This is necessary because the length of a *measured second on our stopwatch* increases at high velocities. Consequently, a larger volume of space now flows per second of our *measured time* [as per our stopwatch]. Therefore, the flowing volume must be multiplied by the amount a measured second increases at this high velocity:

$$(4) \quad e = \frac{F_v}{[L / L_U]} \cdot \frac{1}{\sqrt{1 - V^2/C^2}}$$

[Where: V = Velocity of atom; C = speed of light]

6.4.3. The "Phenomenon of Time" vs. Time

It is very important to reiterate the fact that this proof does not refute the existence of the "phenomenon of time". Rather, to show that time only exists as a function of motion, a phenomenon created by motion, and does not exist upon its own as a fundamental principle of the universe.

Consequently, the use of a "time device" (such as a stopwatch) in this formula is to use a harmonic motion to *measure* the distance between two random motions in exactly the same way as a yardstick is used to measure the distance between two objects.

6.5. Three Important Points of View

One final observation: if fourth dimensional space exists, there are now three points of view from which to observe the 3d universe:

A. The omnipresent point of view from fourth dimensional space. From 4d space, an almost God-like view of 3d space is created: where everything happening in 3d space is seen.
B. The point of view from inside a windowless spaceship:
C. The point of view from inside the electron or the proton.

So, assuming that the hydrogen atom used in this study is part of a spaceship, when the ship begins to move at velocity V, to understand why this atom, and all of the rest of the atoms in the ship change shape, this hydrogen atom must be observed from all three different points of view *simultaneously:*

7. The Mathematics

The proof that time does not exist begins with the spaceship motionless in space [V = 0]. Because V = 0, the velocity of the hydrogen atom [trapped within the skin of the ship] is also V = 0.

7.1. When the Atom is Motionless

When the spaceship is not moving, the hydrogen atom looks as it does in the figure below:

Figure 4

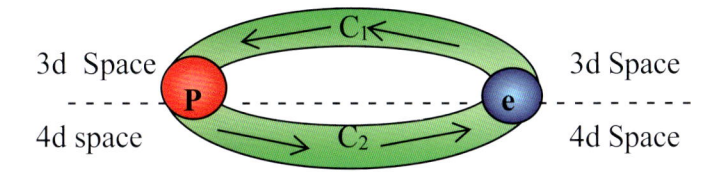

Where: [C1 represents the speed of the 3d vortex]
[C2 represents the speed of the 4d vortex]
(Where C represents the speed of the electrostatic charge = the speed of light)

7.2. The Motionless Atom as Seen From the Three Points of View

The three viewpoints are now used to view the atom when it is motionless in space: V = 0.

A. From the omnipresent point of view, all three of these values are equal:

(5)   [C1 = C2 = C]

B. From the traveler's point of view within the ship, things are different. He cannot see either vortex, nor could he measure the round trip time if he could. The best he can do is measure the round trip time of light between mirrors (perform the Michelson Morley experiment within his spaceship). (And if he does, he finds the speed of light is still equal to "C".)

C. From the point of view within the electron or proton, all three of these values are equal: [C1 = C2 = C] However, when the atom begins to move at velocity V, the situation changes dramatically:

7.3. When the Atom is Moving, the View-point inside the Electron and Proton Change

When the atom is moving at velocity "V", the three viewpoints change:

Figure 5

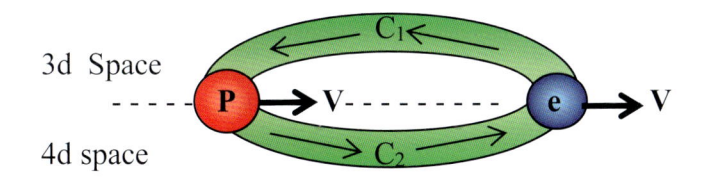

A. From the omnipresent point of view, the velocities of the two vortices are still equal: C1 = C2 = C. Even though everything is moving, <u>***nothing, neither the proton, the electron, nor the flowing vortices are moving faster than the speed of light.***</u>

B. From the traveler's point of view within the spaceship, everything appears normal. The traveler cannot see what is happening within the electron and proton. If he again performs the Michelson Morley experiment, nothing appears to be any different from before. The round trip of the light between the mirrors is still measured to be equal to "C", the speed of light.

C. However, from <u>**INSIDE**</u> the electron or proton, the velocities of the two vortices now "**appear**" to be drastically different:

(6) $\qquad V_{in} = C - V$

(7) $\qquad V_{out} = C + V$

Where: V_{in}=Velocity into electron; V_{out}–Velocity out of electron

7.4. Why There are Three Different Points of View

The reason why the same set of circumstances creates three completely different points of view is explained in the following way:

A. From the omnipresent point of view, the actual speed of the vortices remains the same. Even though the two particles are now moving at velocity "V", **NOTHING IN THIS SYSTEM IS MOVING FASTER THAN THE SPEED OF LIGHT!**

If an imaginary mark could be placed upon the space flowing out of the electron, it would reach the proton quicker, not because it was flowing faster, but rather, because the proton is moving towards it. Even though such a mark arrives at the proton faster than it would if the atom was not moving, the actual velocity of the vortex has not changed: **IT IS STILL FLOWING AT THE SPEED OF LIGHT, "C"**.

The opposite is true for a mark placed upon the space flowing out of the proton. Since the electron is moving away from this mark, as this mark moves towards the electron, it now takes longer to arrive at the electron than it would if the atom was not moving. Even so, it must always be remembered that this vortex is **STILL FLOWING AT THE SPEED OF LIGHT, "C"**.

B. From the traveler's point of view within the spaceship, nothing has changed. Again, since it will be shown that the length of the ship parallel to the direction of travel shrinks, and a measured second slows down, when the traveler again measures the time it takes for the light to travel between the two sets of mirrors he has set up to perform the Michelson Morley experiment, he still sees no difference.

C. HOWEVER, from the point of view within the electron (or proton), things have changed radically. From this viewpoint, the electron (or proton) is **NOT MOVING**. Consequently, the space flowing into the electron now appears to be entering at the **APPARENT VELOCITY** of C-V, and exiting at the **APPARENT VELOCITY** of C+V! (The opposite is true for a viewpoint within the proton.)

7.5 NOTHING IS MOVING FASTER THAN THE SPEED OF LIGHT!!!

And again, it is extremely important to understand that when the atom is moving, **NOTHING, NO PART OF THE ATOM OR ITS VORTICES ARE MOVING FASTER THAN THE SPEED OF LIGHT.** The vortices are still flowing at the speed of light. However, because the positions of the two particles are constantly changing, from the viewpoints **within** the electron and the proton, the velocities of the two vortices are now entering and exiting the particles at different apparent velocities. Hence, the volume of space flowing into a particle no longer equals the volume flowing out of a particle. If nothing changed, an impossible situation would develop where the flow out of a particle would exceed the flow into it. The only way for flowing space to avoid this disparity, is to have some of the volume from the faster flow add to the volume of the slower flow, then, the volume into and out of each particle remains equal. The calculation of this value is designated as the FLOWING VOLUME: F_v

7.6. Calculating the Flowing volume Parallel to the Direction of travel

When the electron and the proton are in the positions seen in Figure 5, they are parallel to the hydrogen atom's velocity of travel (and moving from left to right across the page); the calculation of the FLOWING VOLUME has a simple solution:

Initial conditions: in the following problems, the units are as follows: Speed of light = 1m/sec; Length of vortices = 1m; Electrostatic Charge (e) of electron and proton = 1m^3/sec.
WHERE V = .866C

1. The apparent velocity into the electron is first calculated: from formula **(6)**: $V_{in} = C - V$

$$V_{in} = (1m/sec) - (0.866m/sec) = 0.134m/sec$$

2. The apparent velocity out of the electron is calculated:

From formula **(7)**: $V_{out} = C + V$

$$V_{out} = (1m/sec) + (0.866m/sec) = 1.866m/sec$$

3. To find the FLOWING VOLUME (F_v), the following identity is used:

From formula **(2)**: $V_i = V_o$ [where V_i = volume in; and V_o = volume out]

Since: $V_i = (V_{in} A)$ and $V_o = (V_{out} A)$ [where A = area]

Then: (VELOCITY IN)(AREA) = (VELOCITY OUT)(AREA)

So: (0.134m/sec) (1m^2 + X) = (1.866m/sec) (1m^2 - X)

Solving for X: X = 0.866m^2

(0.134m/sec)(1m^2 + .866m^2) = (1.866m/sec) (1m^2 - 0.866m^2)

.250m^3/sec = .250m^3/sec

F_v = Flowing Volume = .250m^3/sec

Figure 6

Cross sectional views of the two vortices when the atom is moving at 0.866c:

⬅ C + V

Cross sectional area = .134m^2

C–V ➡

Cross sectional area = 1.87m^2

Note how the cross sectional area of the flow traveling at the slower apparent velocity of 1.866c is now about 14 times more than the cross sectional area of the flow traveling at the faster apparent velocity of 0.134c. This now makes the flow into and out of each particle again equal: 0.250m^3/sec = 0.250m^3/sec

However, although they are again equal to each other, they are not equal to the original value of 1m^3/sec. This value is important because it represents the electrostatic charge of the electron. No matter what the velocity of the ship, the traveler inside will always measure the same charge on the electron. Nevertheless, because the value of the charge has now changed, there is a problem: something else must change to restore the electron's charge to its original measured value.

7.7. The Explanation of Length Shrinkage

To restore the electron's charge to its original value the Flowing volume is re-examined:

To determine how the Flowing volume's (F$_v$) value of 0.250m^3/sec changes back to the value of 1m^3/sec, (F$_v$) is substituted into equation **(4)**:

$$e = \frac{[\,0.250 m^3/sec.\,]}{[\,L / L_U\,]} \cdot \frac{1}{\sqrt{1 - V^2/C^2}}$$

Because we know that V = 0.866c, we find that the time dilation effects are equal to:

$$\frac{1}{\sqrt{1 - V^2/C^2}} = \frac{1}{\sqrt{1 - [(0.866 m/sec)/(1 m/sec)]^2}} = 2$$

Since a second on our stopwatch is now twice as long,
e now equals: e = 2[0.250m^3/sec.]   e = .500m^3/sec

However, since the value of e still does not equal its original 1m^3/sec, it is obvious that the length of the radius now has to shrink: but how much? And what causes it to shrink? To understand what causes the atom to shrink, let us recall how a hydrogen atom is created via contemporary science:

When a hydrogen atom is created, the opposite electrostatic charges of the proton and the electron begin to attract the two particles towards each other via Coulomb's law. As they move closer and closer, the distance between them decreases and the Coulomb force between them increases. As the distance continues to decrease, the force continues to increase until the electron reaches the point where it begins to "rotate" about the proton. At this instant, the hydrogen atom is created and the electron is said to be in its ground state. And although we cannot say how the electron "rotates" about the proton or what path it takes about the proton, we do know that the electrostatic force between the proton and the electron has reached a *maximum value.*

However, according to the Vortex Theory, the above description is incomplete. For example, why doesn't the electron keep getting closer and closer? Or when an electron is pushed closer to a proton by the compressed matter in a star, even though the charges are again balanced, why doesn't the electron stay there? Why does it move out to the same radius of all other electrons of all other hydrogen atoms in the universe?

The amazing answer is found in the less dense and the denser regions of space surrounding the proton and the electron respectively.

When the electrostatic charges of the proton and the electron attract these two holes in 3d space together, directly in between them, the less dense space of the proton is slowly canceled out by the denser space surrounding the electron [and vice versa]. This continues to happen until a point is reached where the space directly in-between both holes is balanced. No longer is the space less dense or denser, but is equal throughout, and the Coulomb force between them is neutralized.

If the two holes continue to try to get closer, the density of the space between them now increases, creating an anti-gravity effect, forcing them apart. If they try to increase the distance between them, the now less dense space between them creates a nuclear gravitational effect, pulling them back together. Hence, it is the combination of two forces in balance [the Coulomb force and the nuclear gravitational force] that hold the electron to the proton, and not just one.

However, if the volume of the flow into the proton decreases, the two forces are no longer in balance. Although the charges of the two holes are balanced [the flowing volume of space out of the electron equals the volume of flowing space into the proton]; the space between them is now less dense. The smaller volume of space into the proton has made the surrounding space less dense, while the smaller volume of space surrounding the electron has lessened in density due to the lower outward flow. Consequently, directly in-between the two holes, space is now less dense. This less dense space increases the nuclear gravitational force of the proton, pulling the electron closer - until the space directly in-between them is again equal. As they are pulled together the volume of the space flowing out of one and into the other increases, until again, a point is reached where the two forces are in balance and negate each other.

How much the distance in-between them shrinks – how much the radius of the atom shrinks via the force of nuclear gravity – *can be determined by using Coulomb's law:*

(8) $$F_1 = \frac{k Q_1 Q_2}{d^2}$$

[where F_1 equals the original force between the proton and the electron at "0" velocity] However, we have just seen that the value of the electrostatic charge of both the proton and the electron have dropped from a value of 1, to a value of ½. Consequently,

[when moving at .866 C] : $$F_2 = \frac{k [Q_1 \mathbf{1/2}] [Q_2 \mathbf{1/2}]}{[d(?)]^2}$$

[multiplying the decreases together] :

$$F_2 = \frac{k\, Q_1 [1/2]\, Q_2 [1/2])}{d(?)^2} = \frac{k\, Q_1\, Q_2 [1/4])}{[d(?)]^2}$$

To restore the lower value of F_2 to the normal value of F_1, the amount of shrinkage of the length, [the value of $(?)^2$] can be determined by taking the *square root* of $\sqrt{1/4} = 1/2$. It is now easily seen that the value of the Coulomb Charge, F_2 again equals the value of F_1 when the proton's nuclear gravity causes it to shrink to 1/2 its length!

$$F_2 = \frac{k\, Q_1\, Q_2 [1/4]}{[d(1/2)]^2} = \frac{k\, Q_1\, Q_2 [1/4]}{[d(1/2)]^2} = \frac{k\, Q_1\, Q_2\, \cancel{[1/4]}}{[d]^2\, \cancel{[1/4]}}$$

The numbers [1/4] cancel, restoring the original value of the force to: $F_1 = \dfrac{k\, Q_1\, Q_2}{d^2}$

In conclusion: when the hydrogen atom is traveling at .866c, and the vortices of space flowing between the electron and the proton are parallel to the direction of travel, the charges on the electron and proton are reduced to 1/2 their original value. This causes the atom's radius to shrink by a value of 0.5 to return the value of the space flowing into the proton and out of the electron to a value of $1m^3$. Incredibly, the amount of length shrinkage is **exactly** equal to the amount determined by the Fitzgerald contraction:

$$\sqrt{1 - v^2/c^2} = \sqrt{1 - (.866)^2/(1.000)^2} = 0.5$$

Exactly the same amount of length shrinkage both Lorentz and Fitzgerald proposed a hundred years ago even though they didn't know why or what was causing matter to shrink!

7.8. Why the Radius of the Atom Doesn't Shrink Perpendicular to the Direction of Travel

The atom doesn't shrink in the direction perpendicular to the velocity of travel, because it doesn't have to.

Unlike the direction parallel to the velocity of travel, in the direction perpendicular to the velocity of travel, both flowing volumes have IDENTICAL APPARENT VELOCITIES, making their cross sectional areas equal. Consequently, one flow does not have to subtract from itself to add to the other.

Furthermore, unlike its non-moving condition, the two vortices cannot flow directly towards each particle. When flowing space is discharged from a moving particle towards another particle perpendicular to it, space cannot flow directly towards it. If it did, its resultant vector "R" in Figure 7 would be moving faster than the speed of light "C": a scenario that makes R greater than C; a situation that cannot happen.

Figure 7

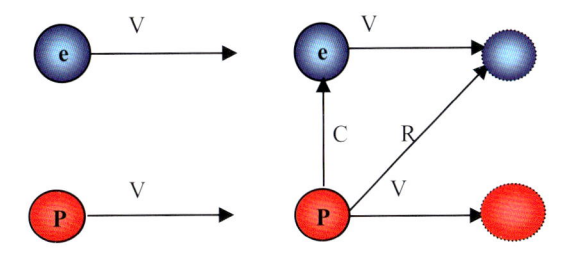

Therefore, when space traveling at the speed of light "C" flows out of one particle towards the other, it now has to travel at the apparent velocity of the "y" vector:

Figure 8

Where C = the speed of light; V = %C = V/C and; y = %C = y/C

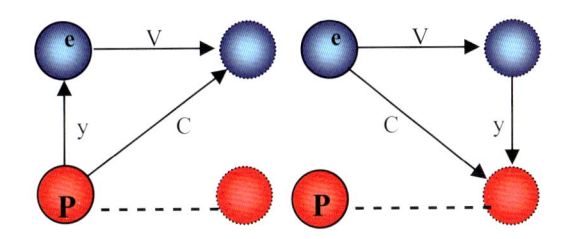

In this position, with both particles directly perpendicular to each other, "y" equals the APPARENT VELOCITY of the space both flowing from the electron to the proton and from the proton to the electron. Because the angles between the "V" vectors and the "C" vectors are the same, both "y" vectors are also equal to each other via similar triangles.

To calculate the value of "y" when v = 0.866c, the Pythagorean Theorem is used:

(9)
$$C^2 = V^2 + y^2$$
$$y^2 = C^2 - V^2$$
$$y^2 = 1^2 - (.866)^2$$
$$y = 0.5 \text{m/sec}$$

Hence:
APPARENT VELOCITY into the electron = 0.5m/sec
APPARENT VELOCITY out of the electron = 0.5m/sec
To calculate the value of the FLOWING VOLUME equation (2) is used:

$$V_i = V_o$$
$$(\text{Velocity In})(\text{Area}) = (\text{Velocity Out})(\text{Area})$$
$$(0.5\text{m/sec})(1\text{m}^2 + X) = (0.5\text{m/sec})(1\text{m}^2 - X)$$
$$(X)(1\text{m/sec}) = 0$$
$$X = 0$$
$$(0.5\text{m/sec})(1\text{m}^2) = (0.5\text{m/sec})(1\text{m}^2)$$
$$0.5\text{m}^3/\text{sec} = 0.5\text{m}^3/\text{sec}$$
$$F_V = \text{Flowing Volume} = 0.5\text{m}^3/\text{sec}$$

To calculate the LENGTH of the flow, Formula **(4)** is used:

$$e = \frac{F_v}{[L/L_U]} \cdot \frac{1}{\sqrt{1 - V^2/C^2}} \qquad [\text{Where } e = 1m^3/sec]$$

Given: $\quad \dfrac{(0.5 m^3/sec)}{L/1m} \cdot \dfrac{1}{\sqrt{1-(0.866)^2}} = 1 m^3/sec$

Since: $\quad \dfrac{(0.5 m^3/sec)}{L/1m} \cdot 2 = 1 m^3/sec$

Then: $\quad \dfrac{1 m^3/sec}{L/1m} = 1 m^3/sec \qquad$ Hence: L = Length = 1m

Amazingly, in the direction perpendicular to the velocity of travel, the charges on the proton and the electron are reduced to one-half of their value. However, because the measured value of a second is also reduced to one-half of its value, twice as much space now flows per-measured second on our slower stopwatch. Consequently, the density of space remains the same. Hence, the proton's nuclear gravity does not increase so it does not pull the electron closer to it, and the radius of the atom does not shrink; *hence, the diameter of the atom does not shrink either: just as predicted by Lorentz and Fitzgerald (even though they did not know why)!*

7.9. The Radius of the Atom From 0^0 to 90^0

The key to discovering the shape of the atom from 0 to 90 degrees is found in the changing lengths of its radius; and the key to discovering the lengths of its changing radius is found in the changing values of the APPARENT VELOCITIES of the two vortices as they rotate with the electron about the proton.

To calculate the APPARENT VELOCITIES of the vortices in the region greater than 0 degrees yet less than 90 degrees, the viewpoints of the omnipresent observer and the traveler in the spaceship must be reexamined.

This viewpoint is limited. We cannot look into the atom and measure the velocities of the vortices. Indeed, we cannot even see these vortices; we can only perceive them with our imaginations.

However, if we could see our world from the viewpoint of the omnipresent observer, we would see a totally different vision of reality. Nothing would seem "normal" to us, especially the directions the vortices flow.

For example, when the electron and proton align to form a 90° angle with the direction of travel, from the travelers' perspective it appears that both vortices are flowing at the speed of light.

Figure 9

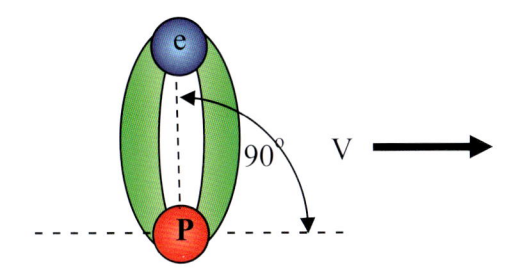

However, as seen from the omnipresent point of view, for a 90° angle the vortices are actually flowing at the slower apparent velocity of the "y" vector in Figure 8.

To understand why this phenomenon is taking place, it is important to reiterate that the traveler is viewing an illusion, while the omnipresent observer is viewing reality: (It is shocking to contemplate the fact that abnormality is creating the semblance of normality!)

The traveler views his surroundings as if everything is sitting dead still, while from 4d space, the omnipresent observer sees the 3d world (or spaceship) in motion. Hence the omnipresent observer sees what is really happening. He sees the atom moving through space at velocity (V) causing the space in the vortices to take longer to flow from the electron to the proton and back. Making it seem as if it is flowing at the slower apparent velocity of the "y" vector.

The reason why the omnipresent observer sees the space flowing at this slower apparent velocity is again due to the fact that while moving at velocity (V), the vortex flowing FROM the proton to the electron cannot flow directly towards it at the speed of light. As previously mentioned, if it did, its resultant vector would exceed the speed of light. Hence, to account for this fact and to create a link between these two completely different visions of the universe, the following formula was developed:

(10) $\quad [(\cos A)(Z_1) - V]^2 + [(\sin A)(Z_1)]^2 = C^2$

This formula is merely an extension of the Pythagorean Theorem with $[(\cos A)(Z_1) - V]^2$ representing the horizontal component, while the value $[(\sin A)(Z_1)]^2$ represents the vertical component. The value of "Z" is the apparent velocity, while "C" is the speed of the vortices: the speed of light.

Since there are two vortices, in formula **(10)** when the cosine is positive, the solution is called the UP VORTEX. The UP VORTEX flows *with* the direction of travel. The UP VORTEX flows through fourth dimensional space from the proton to the electron in Quadrants #1 and #4 of the Cartesian Co-ordinate system when the atom is moving from left to right across the page and the electron is in front of the proton. Because the cosine is positive in Quadrants #1 and #4, Z subtracts from V.

The vortex flowing *back* through three dimensional space from the electron to the proton in Quadrants #1 and #4 is designated the BACK VORTEX. Because the value of the cosine is now negative, Z is negative; making -Z and -V add to each other. This condition arises because the vortex is flowing backwards towards the particle that is moving forward towards it. Making it appear as if this vortex is moving faster. However, it, just like the UP VORTEX is always flowing at "C", the speed of light.

A closer examination of this formula reveals that it also contains the solutions to the two problems we have already solved: the direction PARALLEL to the velocity of travel, and the direction PERPENDICULAR to the velocity of travel:

[When angle A equals 0 degrees, the sine equals 0 while the cosine equals 1; making "Z", the apparent velocity, of the UP VORTEX equal to C - V, and "Z", the apparent velocity of the BACK VORTEX equal to C + V: allowing us to find the length of the atom in the direction PARALLEL to the velocity of travel.

Conversely, when angle A equals 90 degrees, the Sine equals 1, while the Cosine now equals 0. This makes "Z", the apparent velocity of both the UP and BACK VORTICES equal to the square root of C^2 minus V^2: allowing us to find the length of the atom in the direction PERPENDICULAR to the direction of travel].

Note too, that when the atom is not moving, "V" equals 0, making the value of "Z" equal to "C", the speed of light.

To solve for the value of "Z", the quadric formula is used giving us:

(11) $\quad Z = [(V)(\cos A)] \pm \sqrt{(V^2)(\cos A)^2 - V^2 + C^2}$

Note: the addition of the two terms in this equation gives us the value of the apparent velocity for the UP VORTEX, while their subtraction equals the value of the BACK VORTEX. (The difference in their signs indicates they are traveling in opposite directions.)

In summation, once we have found the values of the APPARENT VELOCITY UP and the APPARENT VELOCITY BACK, we can then substitute these values into formula **(2)** and find the value of the FLOWING VOLUME. The value of the flowing volume is then substituted back into formula **(5)** to find the value of the LENGTH OF THE FLOW: which equals the radius of the hydrogen atom at this particular location of the electron on the shell.

FOR EXAMPLE: When the atom is moving at a velocity of .866C, to find the radius of the atom where it makes a 45 degree angle to the direction of the velocity of travel, the following procedure is used:

Step #1 Find the values of the apparent velocities of the Up and Back vortices:

Formula **(11)**

$$Z = [(V)(\cos A)] \pm \sqrt{(V^2)(\cos A)^2 - V^2 + C^2}$$

$$Z = [(.866)(.707)] \pm \sqrt{(.866)^2(.707)^2 - (.866)^2 + (1)^2}$$
$$Z = [0.612] \pm [0.790]$$
$$\text{up} = [0.612] - [0.790] = 0.178 \text{ m/sec}$$
$$\text{Back} = [0.612] + [0.790] = 1.402 \text{ m/sec}$$

[Note, the negative sign only designates direction and is not necessary for our solution.]

Step #2 Find the value of the flowing volume F_v

Formula **(2)**
$$V_i = V_o$$
$$(\text{Vel. In})(\text{Area}) = (\text{Vel. Out})(\text{Area})$$
Since: Velocity Up = Velocity In
And: Velocity Back = Velocity Out
Then: (Velocity Up)(Area) = (Velocity Back)(Area)
So: $(0.178 \text{m/sec})(1\text{m}^2 + X) = (1.402 \text{m/sec})(1\text{m}^2 - X)$
$$X = 0.775 \text{m}^2$$
Hence: $F_v = (0.178 \text{m/sec})(1\text{m}^2 + .775\text{m}^2) = 0.316 \text{m}^3/\text{sec}$ Step #3 To find the length of the radius at 45 degrees to the direction of the velocity of travel:

Formula **(5)**: $\quad e = \dfrac{F_v}{[L/L_U]} \cdot \dfrac{1}{\sqrt{1 - V^2/C^2}}$

$$e = \dfrac{(.316 \text{m}^3/\text{sec})}{L/(1\text{m})} \cdot \dfrac{1}{\sqrt{1 - (.866)^2}} = 1 \text{m}^3/\text{sec}$$

L = Length = .63m

Hence, when the positions of the electron and the proton are such that a line drawn between them forms an angle of 45 degrees to the direction of the velocity of travel, the proton's increased nuclear gravity pulls the electron closer to it until their electrostatic forces are again neutralized: causing the radius of the atom to shrink to a value of .63m.

Plotting a number of angles with their corresponding radii yields the following table:

Table 1

	Values for the + y and + x region		
Angle	Radius	Angle	Radius
0	0.500	50	0.67
5	0.502	55	0.71
10	0.506	60	0.76
15	0.513	65	0.81
20	0.524	70	0.86
25	0.538	75	0.91
30	0.55	80	0.96
35	0.58	85	0.99
40	0.60	90	1.00
45	0.63		

WHEN THE VALUES FOR THE +X, +Y, REGION ARE PLOTTED, THE FOLLOWING SHAPE IS CREATED:

Figure 10

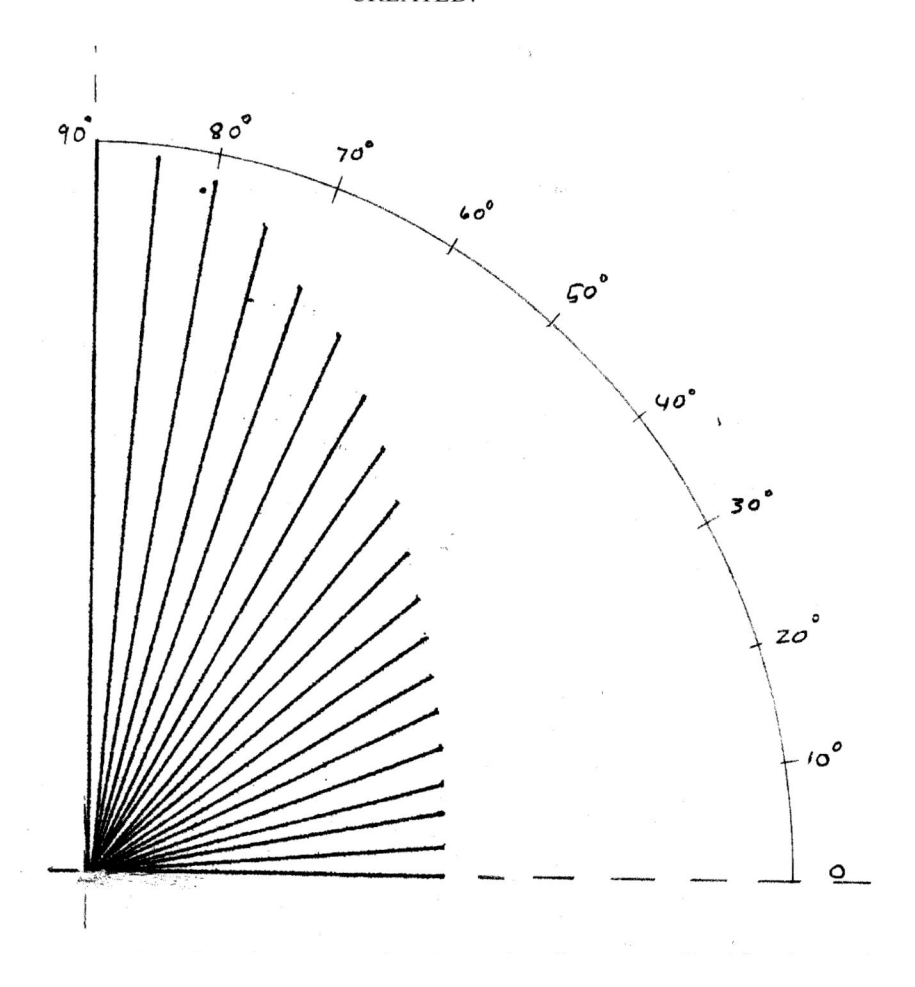

PLOTTING THE VALUES OF ALL FOUR REGIONS, A SPECTACULAR ELLIPTICAL SHAPE COMES INTO VIEW:

Figure 11

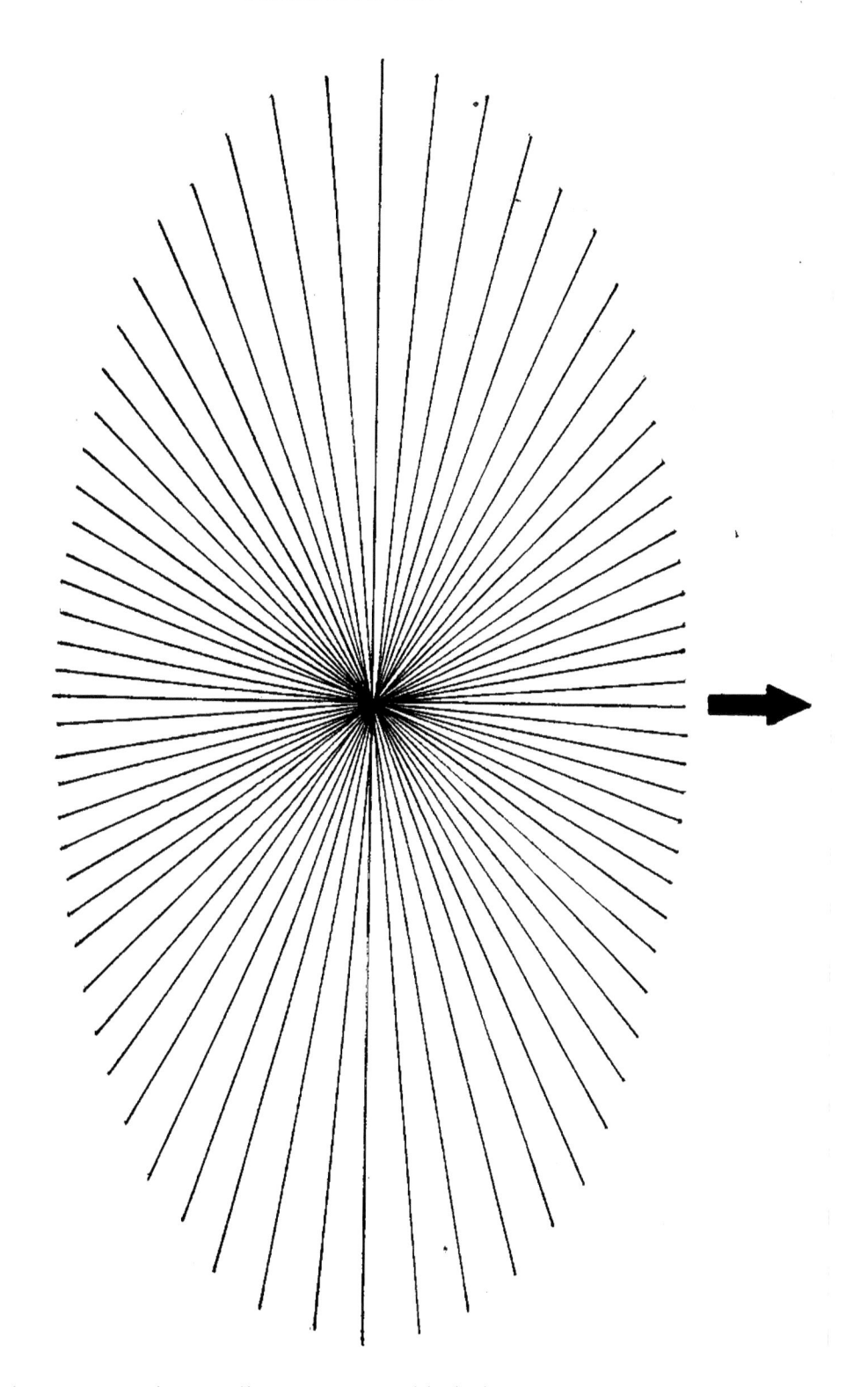

When the hydrogen atom is traveling at .866C, this is its ELLIPTICAL SIDE VIEW; *[from directly in front, it would look like a PERFECT CIRCLE.*

8. Discussion

The elliptical shape of the hydrogen atom moving at .866C not only possesses a kind of rare, geometrical beauty, it also explains time dilation.

When the atom was not moving, according to the perspective inside the electron and the proton, the vortices did not seem to flow at different apparent velocities. Hence, the atom did not shrink; instead, it remained a sphere.

When the atom was not moving, it was spherical shaped. As a sphere, the radius of the atom was the same length for any position of the electron upon its surface. Assuming the radius was 1m, and the speed of light was 1m/s, the time the flowing space took to flow from the proton to the electron and back is 2 seconds. However, this roundtrip time changes when the atom begins to move.

When the atom is moving at V =. 866c, if the length of the radius of the atom, at any point on its shell, is first divided by the apparent velocity of the space in the vortex flowing "up" and then divided by the apparent velocity of the space in the other vortex flowing "back", we discover that the sum of these two different times now equals a round trip time which is *twice* as long.

The following table demonstrates this phenomenon:

Table 2

\multicolumn{7}{c}{[For V = 0.866C]}
Angle A	Radius Length	App. V UP	App. V BACK	Time up	Time Back	Total Time
90 deg	1.0 m	0.5 m/s	0.5 m/s	2 sec	2 sec	4 sec
80	.958	.3718	.6724	2.575	1.424	4.00
70	.860	.2849	.8773	3.019	.980	4.00
60	.756	.2284	1.0944	3.310	.690	4.00
50	.668	.1916	1.305	2.486	.512	4.00
40	.602	.1673	1.494	3.597	.403	4.00
30	.555	.1514	1.6514	3.664	.336	4.00
20	.523	.1410	1.769	3.709	.296	4.00
10	.506	.1358	1.8414	3.725	.275	4.00
0	.500	.1340	1.8660	3.730	.270	4.00

What is the significance of this discovery?

It is not a fourth dimension of "space-time" that is responsible for the creation of time dilation effects in atoms; instead, it is the slower apparent velocity of the vortices themselves.

It is important to understand that we measure "time" <u>using harmonic motions</u> created by devices we call clocks. These harmonic motions allow us to measure the "distance" between shorter random events. Unfortunately, clocks are constructed out of atoms.

Because the clocks in the spaceship are constructed out of atoms, all of the vortices in the clocks will flow at the slower apparent velocity when traveling at velocity "V". In turn, these slower apparent velocities slow the clocks motions down, creating *their* time dilation effects:

9. Time Dilation

It is important to understand that the 3d hole in space we call the electron is itself a creation of the vortex flowing into and out of it. If the electron tried to go faster than the vortex, the space flowing into it from the proton could never reach it. Consequently, its velocity is a function of the velocity of the vortex: dv_e / dv_v [where dv_e = velocity of the electron; and dv_v = velocity of the vortex].

Accordingly, if the speed of the space in the vortices (from its internal point of view) seems to flow slower, the electron must move slower. This deduction comes from the fact that if the electron's speed stayed the same, the length of a second as measured by atomic clocks orbiting the earth in satellites moving at 40,000 km/sec would speed up instead of slow down. However, because atomic clocks slow down in proportion to the Lorentz time dilation formula, we know that the atomic transitions of the electrons in the cesium atom that atomic clocks count slow down according to the slower speed of the measured second. Therefore, when V = .866c, and the round trip speed of the vortices decreases by a factor of .5, the speed of the electron must also decrease by a factor of .5.

Because the speed of the electron about the proton is approximately 1/137c (where c = speed of light), its speed drops to 1/274c; since the proton is at the other end of the vortices, its motions also slow to .5, or one-half their normal speed. Since all the electrons and protons in all of the atoms and molecules in the spaceship experience these same changes, we suddenly realize that all of their motions have slowed down too. This causes all chemical reactions, mechanical reactions, and electrical reactions to slow down. Because all mechanical, electrical, and atomic clocks are made of atoms, clocks will run slower.

Since we already know from the Mathematics of the Michelson Morley experiment that the round trip time of light will slow down between the mirrors, we know that the motions of the photons of energy will slow down via the slower apparent velocity of V: hence, the motions of energy slow down too.

Because men are made of atoms and the atoms within a man are now moving slower, the biological, chemical, and electrical processes within the traveler will take place at a slower rate; hence he will age slower. If he has a twin back upon earth, the "twin paradox" will be created. Although we can go on and on, what we have already seen allows us to begin to realize one of the most profound truth's we have ever encountered in our study of the universe: that everything slowed down because of the slower apparent velocity of V, and not because of a fourth dimension of "space-time".

10. What Creates the Phenomenon of Time?

The phenomenon of time is the result of all the harmonic and sequential atomic, chemical, biological, and astronomical motions creating an orderly sequence of constantly repeating events. These constantly repeating events create a sense of order and harmony in the universe.

However, according to the discoveries in this paper, we can now demonstrate that it is the vortices flowing at the speed of light that are responsible for creating this order and harmony in the universe.

11. The Terminal Velocity of Atoms is the "Speed" of Light!

The maximum speed the atom can move is also a function of the space flowing in the vortex. It is an interesting observation to note that the velocity of the atom itself cannot exceed the speed of the vortex. If it did exceed the speed of the vortex, the vortex flowing in the direction of travel from the proton to the electron in Figure 5 could never catch up to the electron; making the speed of light C, the terminal velocity for all the matter of the universe. But there is something even more amazing:

12. The True Speed of Light Cannot be Measured!

Because of a current misconception that space would have to be very dense to allow light to travel at the incredible speed that it does and therefore could not bend or flow - this erroneous deduction must be dispelled.

The way to correct this erroneous way of thinking is to mention an incredible fact that surfaced during the course of these investigations: the speed of light might not be very fast at all! *In fact, how fast it is really moving is unknown and can never be measured.*

This amazing discovery came from the deduction that every instrument (such as clocks, thermometers, etc.) used as a reference to compare all other motions is made out of atoms. Because the motions of atoms are functions of the speed of the vortices, and since all of the other motions are created by the speed of the vortices, we are using measuring devices whose own motions are themselves functions of the speed of the vortices. These motions are then being used to measure the speeds of other motions that are also functions of the speed of the vortices. Because there are no other motions independent of the vortices to check the speed of the vortices with, the vortices (and the photons of light emitted from them) might very well be flowing at an actual speed of one meter per second or a billion meters per second and we would never know it.

If the vortices were moving at 1m/s, all other motions would slow down accordingly; since the photons of light emitted from the vortices would also flow at 1m/s, everything would continue to appear as it does now. If the speed of the vortices flowed at a hundred billion meters per second, all other motions would speed up as well; the light emitted from the vortices would flow at one hundred billion meters per second - again making everything continue to appear as it does now.

All we can say is this, "That the vortices are moving at what we call 'the speed of light'." We can compare the speed of one motion to another and we can state that one motion is faster or slower than another - but we cannot tell exactly how fast those motions really are because we cannot tell how fast the speed of light really is.

13. The Reconfiguration of Space and the Y Axis

As a hole moves through space, the hole is only moving because the space surrounding it is reconfiguring itself around it. Consequently, the space immediately in front of it has to split apart to get out of the way. If the hole is moving <u>very slowly</u> at velocity dv, the separation takes place at almost the speed of light (C):

Figure 12

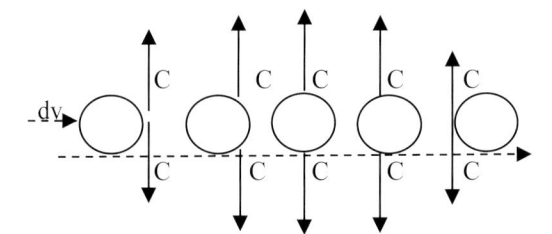

However, if the hole is moving at a relativistic speed of V, such as .866C, space *can no longer reconfigure at the speed of light (C)*. If it did, the resulting vector would exceed the speed of light C, the fastest speed space can move. [Note: if space could move faster, the speed of light would be faster.] Hence, it has to reconfigure at the slower apparent velocity of the Y vector: Y_s [.500C]:

Figure 13

(12) $\quad Y_s \quad$ Where: $Y_s = \sqrt{C^2 - V^2}$
$= \sqrt{C^2 - (.866C)^2} = .500C$

(Where Y_s = velocity of **space**: a % of C.)

In Figure 14 below, we see that at relativistic velocities, the slower velocity of the separation of space now controls the speed of the reconfiguration of the hole as it moves through space.

Figure 14

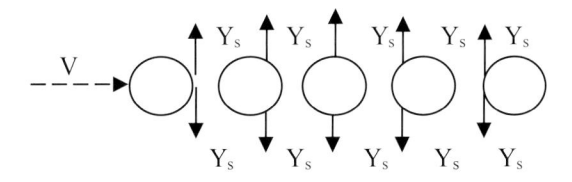

And when the surrounding space reconfigures at the slower apparent velocity of the Y_s vector, the hole cannot deflate as fast as it could when it was not moving. If the above moving hole represents a muon, the muon cannot decay into an electron and two neutrinos as fast as it could when stationary. Since particles moving at velocity V decay at the slower rate of the Lorentz factor:

$dt' = dt / \sqrt{1 - V^2/C^2}$ Since: $\sqrt{1 - V^2/C^2} = \sqrt{C^2 - V^2}$

And from formula **(12)**, we see that: $Y_s^2 = C^2 - V^2$

Substitution gives us: $dt' = dt / \sqrt{Y_s^2}$

Also, because the value Y_s is actually a % of the speed of light C: Y_s / C

Then: **(13)** $dt' = dt / \sqrt{Y_s^2 / C^2}$

From Formula **(12)** we see that Y_s has a domain and range of: $[1C \geq Y_s \geq 0]$;

Formula **(13)** reveals that the slower apparent velocity of the re-configuration of the surrounding space is responsible for the longer lifetimes that muons and other exotic subatomic particles possess when they are accelerated to relativistic velocities.

14. The Slower Apparent Velocity of Space

In this paper, we have seen that when moving at near light velocities, the SLOWER APPARENT VELOCITY of the vortices cause all motions to slow down, creating the phenomenon of "time dilation".

In Table #2 it is shown that for a velocity of .866C, the total round trip time of 4.0 seconds is always twice as long as the previous round trip time of 2.0 seconds when the atom was not moving.

When the round trip times for many different velocities are calculated, the difference between their round trip time and the "at rest" round trip time creates a time dilation effect equal to:

MOVING AT VELOCITY "V"

$$dt' = \frac{dt}{\sqrt{1 - V^2/C^2}} = \text{TIME DILATION EFFECTS}$$

Where dt' equals the difference in a measured second of time in the moving frame of reference traveling at velocity "V", and the measured second of time (dt) in the non-moving frame of reference.

However, for subatomic particles moving at near light velocities, the slower reconfiguration of space moving at the Y_s vector creates the exact same effect; causing them to decay at a slower rate. And when many values of Y_s are plotted, it is discovered that the graph for the time dilation effects [Figure 15] created by the slower velocity of reconfiguring space, is <u>identical</u> to the one made by using Lorentz's time dilation equation [Figure 16]:

Figure 15 $$dt' = dt/\sqrt{Y_s^2/C^2}$$

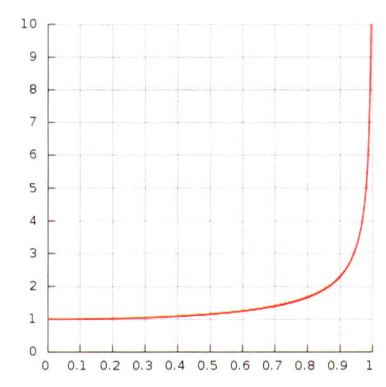

[dt'/dt is vertical; V/C is horizontal]

Figure 16 $$dt' = dt\sqrt{1 - V^2/C^2}$$

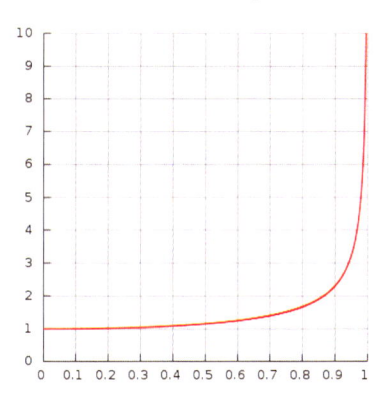

[dt'/dt is vertical; V/C is horizontal]

But even more amazingly, without ever having to study the Theory of Relativity, the discovery that slower reconfiguring space can create time dilation effects, allows us to next deduce that time dilation effects can also be created by the inability of space to reconfigure itself at the speed of light in other regions of the universe: such as those within powerful gravitational fields.

14.1. Time Dilation Formula vs. Space Dilation Formula for Gravitational Fields

If a situation develops in some area of the universe where space can no longer move at its maximum rate, the motions of atoms in this area will slow down, creating greater distances between the repetition of events occurring here, and the repetition of similar events occurring elsewhere. Such a situation would make it appear as if "time" has somehow slowed down, which is precisely what happens within intense gravitational fields.

In the powerful gravitational fields surrounding stars, space is less dense and cannot reconfigure around a hole in space as fast as it can where space is denser: such as in the region between the Galaxies.

Where space possesses its <u>maximum density</u> is in the vast regions between the Galaxies where there is almost an absence of protons, electrons, and neutrons. Expressed mathematically, we can call this dense region of space d_E: "d" for density; "E" for elasticity.

Let us next assign d_E a value of "C" [here, space can move at C, or 300,000 m/s]. In contrast, in all other <u>less dense</u> regions of space d_E', will be percentages of "C" [d_E' = % of C; giving d_E' a domain and range existing between: $C \geq d_E' > 0$].

[Note: d_E' represents the less dense region of space in a gravitational field. Although the density of space is at its least where gravitation is at its highest, (such as within a Black Hole), nevertheless, it can *never reach the value of "0"*. If it did, the space surrounding the black hole could never reconfigure and it could never move; also, as the black hole absorbed other matter [holes] its surface could never bend and flex; hence, it could never expand in size: or eventually deflate. Consequently, it must possess some flexibility; giving it a <u>minimum density</u> value of $d_E' > 0$.]

However, even though we cannot, at this present moment in the history of science, define exactly what the maximum and minimum parameters are, we can still construct the formula for gravitational time dilation. We begin by dividing d_E' by d_E:

This division gives us: [d_E' / d_E]

Next, to equate the differences in time dilation between these maximum and minimum densities of space, we use the following formula:

$$dt' = dt / [\ ?\ ]$$

And then substituting d_E' / d_E, we get:

(14) $dt' = dt / [d_E' / d_E]$

In the above formula, the *decrease* in the elasticity of space in a gravitational field makes it harder to bend and flex. As such, this phenomenon is then responsible for the slower reconfiguration of space surrounding a hole in space or surrounding a fast moving dense region of space such as a photon.

This change in the "elasticity" of space explains why intense gravitational fields create time dilation effects. According to the Vortex Theory, the gravitational fields surrounding all astrological bodies are created by the addition of all the less dense regions of space surrounding all the protons and neutrons in the astrological body.

The intense "gravitational fields" created by the massive regions of less dense space surrounding stars, changes the elasticity of space - making it move slower – making the motions of everything move slower: creating time dilation effects. Answering the last of the enigmas of Relativity by allowing us to understand why "gravity" affects the measurement of time.

15. The Speed of Light is a Variable and Only Appears to be a Constant

The speed of light increases in dense regions of space, such as that between the Galaxies, and decreases in less dense regions, such as those within gravitational fields.

The reason why this phenomenon is not seen by the observer is due to the fact that the atoms and molecules the observer's instruments [clocks] are constructed out of also slow down or speed up. Consequently, in a less dense region of space, such as within the gravitational field of the Earth, the speed of light slows down; but the clocks used to measure the speed have also slowed down, making it seem as if no change has been made.

Of course, when the observer is moving, length shrinkage effects come into play and again, no change is noted in the measurement of the speed of light.

16. The End of the Era of Relativity

The era of the Theory of Relativity comes to an end when it is realized that Albert Einstein's vision of the universe is based upon the effects we see and not upon the cause of these effects.

From Einstein's point of view, the motions of everything in the universe were "relative" to the motions of everything else. This relationship is not the true reality; nevertheless, it is a real effect. Hence, the Lorentz transformation equations that allow us to calculate the "time differences" between two moving frames of reference are still valid. In addition, even though the "twin's paradox", the orbit of Mercury aberrations, and many other observations from the "relativistic" vision of the universe are still real effects, the causes of these effects have nothing to do with a fourth dimension of "space-time".

Although it is true that the space surrounding the sun and other large interstellar objects appears to be bent, it is not the "bend" that is responsible for the creation of the force of gravity. The "bend" is a phenomenon created by the less dense region of space surrounding the sun. This less dense region of space is responsible for the creation of the force of gravity and for the bending of starlight seen during eclipses.

Although many ideas regarding time and space have to be discarded or amended, many mathematical formulas are still usable. Just as Newton's laws are still applicable even though the Theory of Relativity amended them, the Theory of Relativity is still applicable even though the Vortex Theory amends it, because the Theory of Relativity is a real phenomenon.

However, in the "microscopic" world inside the atom, things are very different. The view of the universe from inside a proton or an electron is completely different from the relativistic view. Completely different too is our vision of time.

According to relativity, time exists along with matter, space, energy and the forces of nature as one of the five fundamental "pieces" of the universe. Furthermore, according to Albert Einstein, time exists as part of a fourth dimension of the universe called "space-time". However, this idea is and always was conjecture. No fourth dimension of time has ever been discovered.

Also, according to the relativistic vision of the universe, time flowed at the speed of light: the maximum speed possible in the universe. Consequently, because any velocity "V" of matter cannot add to the speed of time, the velocity "V" of matter creates time dilation effects. But this too is a mistake. According to the discoveries in this paper, we can now demonstrate that it is the vortices themselves flowing at the speed of light that are responsible for creating the phenomenon of time; and it is the slower *apparent* velocity of the vortices [still flowing at the speed of light] that are responsible for creating the phenomenon of time dilation.

Although the idea that time does not exist seems offensive to some, this unpleasant emotion can be dispelled by understanding that modern peoples did not invent time. The concept of time is so old it predates writing. It appears to be invented by ancient peoples during their efforts to keep track of the seasons in order to plant crops. But no matter what the reason, the fact still remains that time was invented by people totally ignorant of the construction of the physical universe. People who did not even know the world was round.

Another important misunderstanding that must be cleared up regards clocks. Although some people believe that clocks keep track of time, this too is a mistake. Clocks are only associated with time. Clocks keep track of a position on the rotating surface of the earth in relation to the sun when it is directly overhead.

The length of an hour, minute, and second is arbitrary. They are only useful upon this planet. When men go to Mars, earthly clocks will no longer allow the user to predict when the sun will be directly overhead. Because Mars rotates approximately one-half hour longer than the earth, in twenty-four Martian days, when the earthly clock indicates it is noon, it will really be midnight. Hence, the length of the hour, minute, and second will have to be modified or scrapped altogether on Mars.

[Note: the above thesis was published in 2012 by the Russian Academy of Sciences, St. Petersburg State University Branch; Saint Petersburg, Russia. See **Peer Review Scientific Paper References**.]

17. In Conclusion

This paper has revealed the fact that one does not need to have a fourth dimension of "space-time" to create time dilation or length shrinkage. The construction of space and atoms is the key to understanding what is really happening. When we understand that space is made of something and particles of matter are holes in its surface, the explanations of time dilation and length shrinkage are easily explained.

Is there physical proof?

If this theory is true, the electron has to possess an anti-gravity field. For example, earlier in this presentation, Figure 2 showed that space was flowing into the proton and out of the electron. The flow, or rather the pull of space going into the proton is responsible for creating its gravitational field and the force of gravity. Consequently, if space is indeed flowing out of the electron it is then pushing the surrounding space outward, creating an anti-gravity field.

If this "Anti-gravity" field surrounding the electron can be found, it will not only prove that this Vortex Theory is true, but it will reveal that fantastic new anti-gravity technologies are just waiting to be discovered!

And after eight years of searching, in November of 2013, this is exactly what happened. The anti-gravity field surrounding the electron was discovered! This revolutionary, historical discovery was published in a landmark paper concurrently with this one in 2012 by the Branch of the Russian Academy of Sciences in St. Petersburg State University, Russia: see References below…

References

National/International Conferences attended, and peer reviewed scientific papers presented

[1] The Vortex Theory of Matter. [Presentation of his own work]
'International Forum on New Science' Colorado State University (1992, Sept 17-20).
Moon. R. Fort Collins, Colorado. USA. Topic: The Vortex Theory of Matter. Copyright 1990)

[2] The Vortex Theory and some interactions in Nuclear Physics. [Book of abstracts; p. 259]
'The LIV International Meeting on Nuclear Spectroscopy and Nuclear Structure; Nucleus 2004' (2004, June 22-25). Moon, R., Vasiliev, V. Belgorod, Russia.
http://nuclpc1.phys.spbu.ru/nucl/Abstracts/Nucleus_2004.pdf

[3] The Possible Existence of a new particle: The Neutral Pentaquark. [Book of materials; pp. 98-104]
'Scientific Seminar of Ecology and Space' (2005, February 22). Scientific Research Centre for Ecological Safety of the Russian Academy of Sciences. Moon, R. Saint Petersburg, Russia.
https://spcras.ru/ensrcesras/

[4] Explanation of Conservation of Lepton Number. [Book of materials; p. 105]
'Scientific Seminar of Ecology and Space' (2005, February 22). Scientific Research Centre for Ecological Safety of the Russian Academy of Sciences: Moon, R., Vasiliev, V. Saint Petersburg, Russia.
https://spcras.ru/en/srcesras/

[5] Explanation of Conservation of Lepton Number. [Book of abstracts; p. 347]
'LV National Conference on Nuclear Physics' (2005, June 28-July 1). FRONTIERS IN THE PHYSICS OF NUCLEUS. Moon, R., Vasiliev, V. Russian Academy of Sciences. St. Petersburg State University. Saint Petersburg, Russia.
http://nuclpc1.phys.spbu.ru/nucl/Abstracts/Frontiers_2005.pdf

[6] The Possible Existence of a New Particle: the Tunneling Pion. [Book of abstracts; p. 348]
'LV National Conference on Nuclear Physics' (2005, June 28-July 1). FRONTIERS IN THE PHYSICS OF NUCLEUS. Moon, R., Vasiliev, V. Russian Academy of Sciences. St. Petersburg State University. Saint Petersburg, Russia.
http://nuclpc1.phys.spbu.ru/nucl/Abstracts/Frontiers_2005.pdf

[7] The Possible Existence of a New Particle in Nature: the Neutral Pentaquark. [Book of abstracts; p. 349] 'LV National Conference on Nuclear Physics' (2005, June 28-July 1). FRONTIERS IN THE PHYSICS OF NUCLEUS. Vasiliev, V. Moon, R. Russian Academy of Sciences. St. Petersburg State University. Saint Petersburg, Russia.
http://nuclpc1.phys.spbu.ru/nucl/Abstracts/Frontiers_2005.pdf

[8] The Experiment that discovered the Photon Acceleration Effect. [Book of abstracts; p. 77]
'International Symposium on Origin of Matter and the Evolution of Galaxies' (2005, Nov 8-11). Gridnev, K., Moon, R., Vasiliev, V. New Horizon of Nuclear Astrophysics and Cosmology. University of Tokyo, Japan.
https://meetings.aps.org/Meeting/SES05/Content/273
https://flux.aps.org/meetings/bapsfiles/ses05_program.pdf

[9] The Conservation of Lepton Number. [Paper presentation]
'American Physical Society 72nd Annual Meeting of the Southeastern Section of the APS' (2005, Nov 10-12). Moon, R., Calvo, F., Vasiliev, V. Gainesville, FL. USA. APS Session BC Theoretical Physics I, BC 0008
https://meetings.aps.org/Meeting/SES05/Content/273
https://flux.aps.org/meetings/bapsfiles/ses05_program.pdf

[10] The Vortex Theory and the Photon Acceleration Effect. [Paper presentation]
'American Physical Society; March Meeting; Topics in Quantum Foundations' (2006, March 13-17). Gridnev, K., Moon, R., Vasiliev, V. Baltimore, Maryland. USA.
Abstract ID: BAPS.2006.Mar.B40.6
https://meetings.aps.org/Meeting/MAR06/Session/B40.6
http://meetings.aps.org/link/BAPS.2006.MAR.B40.6

[11] The St Petersburg State University experiment that discovered the Photon Acceleration Effect.
'American Physical Society; March Meeting' GENERAL POSTER SESSION (2006, March 13-17).Gridnev, K., Moon, R., Vasiliev, V. Baltimore, Maryland. USA.
Abstract ID: BAPS.2006.MAR.Q1.146
https://meetings.aps.org/Meeting/MAR06/Session/Q1.146
http://meetings.aps.org/link/BAPS.2006.MAR.Q1.146

[12] The Neutral Pentaquark.
'American Physical Society; March Meeting' GENERAL POSTER SESSION (2006, March 13-17).Moon, R., Calvo, F., Vasiliev, V. Baltimore, Maryland. USA.
Abstract ID: BAPS.2006.MAR.Q1.147
https://meetings.aps.org/Meeting/MAR06/Session/Q1.147
http://meetings.aps.org/link/BAPS.2006.MAR.Q1.147

[13] The Neutral Pentaquark. [Paper presentation]
'International Workshop on "Nuclear Physics with RIBF' (2006, March 13-17).
Vasiliev, V., Calvo, F., Moon, R. RIKEN Research Institution. Saitama, JAPAN.
Abstract: RIBF-Pentaquark.
https://ribf.riken.jp/RIBF2006/

[14] Nuclear Structure and the Vortex Theory. [Paper presentation]
'International Workshop on "Nuclear Physics with RIBF' (2006, March 13-17).
Moon, R., Vasiliev, V. R. RIKEN Research Institution. Saitama, JAPAN.
Abstract RIBF-Vortex
https://ribf.riken.jp/RIBF2006/

[15] Experiment that Discovered the Photon Acceleration Effect. [Paper presentation]
'International Workshop on "Nuclear Physics with RIBF' (2006, March 13-17).
Moon, R., Vasiliev, V. R. RIKEN Research Institution. Saitama, JAPAN.
Abstract Moon 1
https://ribf.riken.jp/RIBF2006/

[16] To the Photon Acceleration Effect. [Paper presentation]
'APS/AAPT/SPS Joint Spring Meeting' (2006, March 21-23).
Moon, R. San Angelo, Texas. USA. Abstract ID: BAPS.2006.TSS.POS.8
https://meetings.aps.org/Meeting/TSS06/Session/POS.8
http://meetings.aps.org/link/BAPS.2006.TSS.POS.8

[17] The Saint Petersburg State University Experiment that discovered the Photon Acceleration Effect. [Paper presentation] 'American Physical Society; Astroparticle Physics II' (2006, April 22-25).
Gridnev, K., Moon, R., Vasiliev, V. Dallas, TX. USA. Abstract ID: BAPS.2006.APR.J7.6
https://meetings.aps.org/Meeting/APR06/Session/J7.6
http://meetings.aps.org/link/BAPS.2006.APR.J7.6

[18] The Photon Acceleration Effect. [Paper presentation]
'American Physical Society; Session W9 DNP: Nuclear Theory II' (2006, April 22-25).
Gridnev, K., Moon, R., Vasiliev, V. Dallas, TX. USA. Abstract ID: BAPS.2006.APR.W9.6
https://meetings.aps.org/Meeting/APR06/Session/W9.6
http://meetings.aps.org/link/BAPS.2006.APR.W9.6

[19] The Neutral Pentaquark. [Paper presentation]
'American Physical Society; Session W9 DNP: Nuclear Theory II' (2006, April 22-25).
Moon, R., Calvo, F., Vasiliev, V. Dallas, Texas. USA. Abstract ID: BAPS.2006.APR.W9.9
https://meetings.aps.org/Meeting/APR06/Session/W9.9
http://meetings.aps.org/link/BAPS.2006.APR.W9.9

[20] Controversy surrounding the Experiment conducted to prove the Vortex Theory. [Paper presentation] 'American Physical Society; 8th Annual Meeting of the Northwest Section' (2006, May 18-20). Vasiliev, V., Moon, R. University of Puget Sound. Tacoma, Washington. USA.
Abstract ID: BAPS.2006.NWS.C1.9
https://meetings.aps.org/Meeting/NWS06/Content/518
https://meetings.aps.org/Meeting/NWS06/Session/C1.9

[21] The Photon Acceleration Effect. [Paper presentation]
'American Physical Society; 8th Annual Meeting of the Northwest Section' (2006, May 18-20).
Moon, R., Vasiliev, V. University of Puget Sound. Tacoma, Washington. USA.
Abstract ID: BAPS.2006.NWS.C1.8
https://meetings.aps.org/Meeting/NWS06/Content/518
https://meetings.aps.org/Meeting/NWS06/Session/C1.8
http://meetings.aps.org/link/BAPS.2006.NWS.C1.8

[22] Experiment that Discovered the Photon Acceleration Effect. [Paper presentation]
'International Congress on Advances in Nuclear Power Plants' ICAPP '06, (2006, June 4-8).
Gridnev, K., Moon, R. Reno, Nevada. USA. American Nuclear Society.
Abstract 6006. ISBN: 978-0-89448-698-2

[23] The Neutral Pentaquark. [Paper presentation]
'International Congress on Advances in Nuclear Power Plants' ICAPP '06 (2006, June 4-8).
Vasiliev, V., Calvo, F., Moon, R. Reno, Nevada. USA. American Nuclear Society.
Abstract 6045. ISBN: 978-0-89448-698-2

[24] Is Hideki Yukawa's explanation of the strong force correct?
'The International Symposium on Exotic Nuclei' Book of abstracts: Joint Institute for Nuclear Research. (2006, July 17-22). Vasiliev, V., Moon, R. Khanty Mansiysk, Siberia. Russia.
http://wwwinfo.jinr.ru/exon2006/
http://jinr.ru/

[25] The Explanation of the Pauli Exclusion Principle. [Paper presentation]
'59th Annual meeting of the American Physical Society Division of Fluid Dynamics' (2006, Nov 19-21). Moon, R., Vasiliev, V. Tampa, Florida. USA. American Physical Society;
Abstract ID: BAPS.2006.DFD.P1.17
https://meetings.aps.org/Meeting/DFD06/Content/578
https://meetings.aps.org/Meeting/DFD06/Session/P1.17
http://meetings.aps.org/link/BAPS.2006.DFD.P1.17

[26] Is Hideki Yukawa's explanation of the strong force correct? [Paper presentation]
'59th Annual meeting of the American Physical Society Division of Fluid Dynamics' (2006, Nov 19-21). Moon, R., Vasiliev, V. Tampa, Florida. USA. American Physical Society;
Abstract ID: BAPS.2006.DFD.P19
https://meetings.aps.org/Meeting/DFD06/Content/578
https://meetings.aps.org/Meeting/DFD06/Session/P1.19
http://meetings.aps.org/link/BAPS.2006.DFD.P1.19

[27] The Final Proof of the Michelson Morley Experiment; The explanation of Length Shrinkage and Time Dilation. [Book of materials] 'Scientific Research Center for Ecological Safety of the Russian Academy of Sciences: Scientific Seminar of Ecology and Space'. (2007, February 8-10). Moon, R. Saint Petersburg, Russia.
https://spcras.ru/en/srcesras/

[28] The Explanation of the Photon's Electric and Magnetic fields and its Particle and Wave Characteristics. [Paper presentation] 'Annual Meeting of the Division of Nuclear Physics Volume 52, Number 10'. (2007, Oct 10-13). Moon, R., Vasiliev, V. Newport News, Virginia. USA. American Physical Society; Abstract ID: BAPS.2007.DNP.BF.15
https://meetings.aps.org/Meeting/DNP07/Session/BF.15
http://meetings.aps.org/Meeting/DNP07
http://meetings.aps.org/link/BAPS.2007.DNP.BF.15

[29] The St. Petersburg State University experiment that discovered the Photon Acceleration Effect. 'Virtual Conference on Nanoscale Science and Technology' VC-NST. (2007, Oct 21-25). Moon, R., Vasiliev, V. University of Arkansas. 222 Physics Building. Fayetteville, AR 72701 USA.
http://www.ibiblio.org/oahost/nst/index.html

[30] The Explanation of Quantum Teleportation and Entanglement Swapping. [Paper presentation] '49th Annual Meeting of the Division of Plasma Physics, Volume 52, Number 11' (2007, Nov 12–16). Moon, R., Vasiliev, V. Orlando, Florida. American Physical Society;
Abstract ID: BAPS.2007.DPP.UP8.21
https://meetings.aps.org/Meeting/DPP07/Content/901
http://meetings.aps.org/link/BAPS.2007.DPP.UP8.21
https://meetings.aps.org/Meeting/DPP07/Session/UP8.21

[31] The Explanation of the Photon's electric and magnetic fields, and its particle and wave characteristics. [Paper presentation]
'60th Annual Meeting of the Division of Fluid Dynamics'. Volume 52, Number 12. (2007, Nov 18–20). Moon, R., Vasiliev, V. Salt Lake City, Utah. American Physical Society;
Abstract ID: BAPS.2007.DFD.JU.22
http://meetings.aps.org/Meeting/DFD07
https://meetings.aps.org/Meeting/DFD07/Session/JU.22
http://meetings.aps.org/link/BAPS.2007.DFD.JU.22

[32] The Explanation of quantum entanglement and entanglement swapping. [Poster Session]
'The 10th International Symposium on the Origin of Matter and the Evolution of the Galaxies (OMEG07) (2007, Dec 4-6) Moon, R., Vasiliev, V. Hokkaido University, Sapporo, Japan. Bibcode: 2008AIPC.1016.....S, Harvard (Astrophysics Data System) ISBN 0735405379
https://ui.adsabs.harvard.edu/abs/2008AIPC.1016.....S/abstract

Books by author {A} and work presented in other published books/booklets

1. *"The Vortex Theory of Matter"* Copyright 1990
   R. Moon. {A} Costa Mesa, California

2. *"The End of The Concept of Time"* Copyright 2000.
   R. Moon. {A} Gordon's Publications of Baton Rouge. Louisiana. ISBN 096792981-4.

3. *"The Bases of the Vortex Theory of Space"* (2002).
   R. Moon. {A} Publishing house; "ZNACK" Director Dr. I. S. Slutskin. Post Office Box 648. Moscow, 101000, Russia. p. 32. (In Russian). Journal ISSN: 2362945.

4. *"The Vortex Theory…The Beginning"* (2003). Copyright 2003.
   R. Moon. {A} (Editor's note by Prof., Dr. Victor V. Vasiliev)
   Gordon's Publications of Fort Lauderdale Fla. USA.

5. *"The Bases of the Vortex Theory"* (2003).
   Book of abstracts: Russian Academy of Sciences; ISBN 5-98340-004-5; TRN: RU0403918096768 OSTI ID: 20530263 R. p. 251. R. Moon. V. Vasiliev
   http://nuclpc1.phys.spbu.ru/nucl/Abstracts/Nucleus_2003.pdf
   http://physics.doi-vt1053.com/ISBN5-98340-004-5/Nucleus_2003.pdf

6. *"The Vortex Theory and some interactions in Nuclear Physics"* (2004).
   Book of abstracts: Russian Academy of Sciences; ISBN 5-9571-0075-7 p. 259.
   R. Moon. V. Vasiliev
   http://nuclpc1.phys.spbu.ru/nucl/Abstracts/Nucleus_2004.pdf
   http://physics.doi-vt1053.com/ISBN5-9571-0075-7/Nucleus_2004.pdf

7. The Vortex Theory Explains the Quark Theory. (2005).
   R. Moon. {A} Gordon's Publications of Fort Lauderdale, Florida. USA. p. 205.

8. Dr. Russell Moon PhD Thesis; *"The End of "Time"* Collection of Learned Works Addendum, 2012, (pp. 473-488) VVM Publishing House: ISBN 978-5-9651-0804-6 Editor in Chief: I. S. Ivlev. Saint Petersburg State University. St Petersburg, Russia.
   http://physics.doi-vt1053.com/ISBN978-5-9651-0804-6/Dr-Russell-G-Moon-PhD-thesis-The-End-of-Time.pdf
   http://physics.doi-vt1053.com/ISBN978-5-9651-0804-6/Natural_Anthropogenic_Aerosoles_4pages.pdf

9. *"The Discovery of the Fifth Force in Nature: The Anti-Gravity Force"* Collection of Learned Works (pp. 489-495) R. Moon. M. F. Calvo.
   VVM Publishing House: ISBN 978-5-9651-0804-6 p. 534. Editor in Chief: I. S. Ivlev. St Petersburg State University. St Petersburg, Russia.
   http://physics.doi-vt1053.com/ISBN978-5-9651-0804-6/The-Discovery-of-the-Fifth-Force-in-Nature:-The-Anti-gravity-Force.pdf
   http://physics.doi-vt1053.com/ISBN978-5-9651-0804-6/Natural_Anthropogenic_Aerosoles_4pages.pdf

10. *"The Discovery of the Fifth Force in Nature: The Anti-gravity Force"* Collection of Learned Works (pages 496-503) V. Vasiliev. R. Moon. M. F. Calvo. VVM Publishing House; ISBN 978-5-9651-0804-6 2013. p 534. Editor-in-Chief: I. S. Ivlev, Saint Petersburg State University. St. Petersburg. Russia.
http://physics.doi-vt1053.com/ISBN978-5-9651-0804-6/The-Discovery-of-the-Fifth-Force-in-Nature:-The-Anti-gravity-Force.pdf

Other References

1) Christopher Scarre. Smithsonian Institution; '*Smithsonian Timelines of the Ancient World'*. p 65. Published September 15th 1993. ISBN-10: 1564583058

2) H. Yukawa, "Tabibito" (The traveler) World Scientific, Singapore. (1982) pp. 190-202. ISBN-10 9971950103

3) Wolfgang Pauli, Nobel Lecture; *for the discovery of the exclusion principle.* Stockholm, Sweden, (1946).
https://www.nobelprize.org/uploads/2018/06/pauli-lecture.pdf

4) Robert Desbrandes, Daniel Van Gent, '*Intercontinental quantum liaisons between entangled electrons in ion traps of thermoluminescent crystals'*. (2006-11-09), arXiv:quant-ph/0611109.
https://arxiv.org/ftp/quant-ph/papers/0611/0611109.pdf
https://doi.org/10.48550/arXiv.quant-ph/0611109

5) D. Bouwmeester, J.-W. Pan, K. Mattle, M. Eibl, H. Weinfurter, A. Zeilinger, '*Experimental Quantum Teleportation',* Nature 390, 6660, 575-579 (1997), arXiv:1901.11004.
https://doi.org/10.48550/arXiv.1901.11004

6) Einstein, A. and Podolsky, B. and Rosen, N., '*Can Quantum-Mechanical Description of Physical Reality Be Considered Complete?'* Phys. Rev. 47, 777, (1935).
https://link.aps.org/doi/10.1103/PhysRev.47.777
https://journals.aps.org/pr/abstract/10.1103/PhysRev.47.777

7) Rucker, Rudy. '*The Fourth Dimension: Toward a Geometry of Higher Reality'*. (1984) ISBN-10: 0486779785

8) Abbott, Edwin, A. *FLATLAND*: '*A Romance of Many Dimensions'*. New York, Dover Publications. (1953) ISBN-13: 9798630248015

9) Besancon, Robert M. '*The Encyclopedia of Physics'*. pp 568-570; and 949-950. Van Nostrand Reinhold Co. (1974) ISBN-10: 1124004475

10) Condon, E. U. '*The Handbook of Physics Second Edition'*. McGraw Hill Book Co. (1967) ISBN-10: 0070124035

11) Eisberg, Robert. '*Fundamentals of Modern Physics'*. Pages 9-15. John Wiley and Sons. (1961) ISBN-10: 047123463X

12) Dorothy Michelson Livingston. '*The Master of Light'*. A biography of Albert A. Michelson. University of Chicago Press. (1973) ISBN-10: 0684134438

Russian Scientific Journals:

1. http://www.new-philosophy.narod.ru/RGM-VVV-RU.htm (in Russian)

2. http://www.new-philosophy.narod.ru/MV-2003.htm (in English)

3. http://www.new-idea.narod.ru/ivte.htm (in English)

4. http://www.new-idea.narod.ru/ivtr.htm (in Russian)

5. http://www.new-philosophy.narod.ru/mm.htm (in Russian)

Some Subjects found in Book 2:

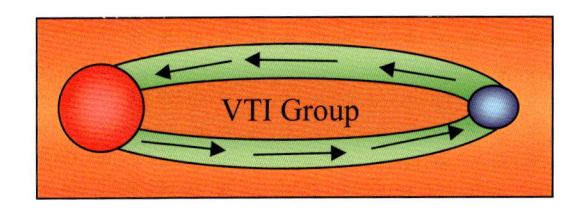

PART I THE TRUE VISION OF THE UNIVERSE

- The Mystery Amid the Foundations of Science
- The Problem is Time!
- The Undiscovered Territory
- The True Vision of Space
- The Proton and the Electron
- The Vortex
- The Neutron
- Particle Collisions
- The True Vision of Energy
- Creating the Forces of Nature
- The Force of Gravity Explained
- Explaining the Mystery of Mass
- The Electromagnetic Force Explained
- The Weak Force Explained
- The Strong Force Explained
- The Anti-gravity Force Explained
- The Shocking Truth About Time
- Time Dilation is Finally Explained
- The True Vision of the Universe
- The Mathematics of the Michelson Morley Experiment
- The Phenomenon of Time Explained
- Time Dilation Explained

PART II SCIENTIFIC DISCOVERIES FROM BOOK 2

- The Explanation of the Particle and Wave Theory of Light
- The Explanation of the Particle and Wave Theory of Matter
- The Explanation of The Double Slit Interference Patterns
- The Explanation of Intrinsic Spin [1/2 spin]
- The Explanation of Newton's Three Laws of Motion
- The Explanation of the Conservation of Charge
- The Explanation of the Conservation of Angular Momentum
- Explaining the Conservation of Momentum
- The Conservation of Mass and Energy
- The Explanation of the Mystery of Entropy
- The Explanation of the Dark Energy and Dark Matter
- Anti-gravity Engineering?
- What the Neutrino Really is!
- The Explanation of Buoyancy
- The Explanation of Covalent and Ionic Bonds in Chemistry
- The Explanation of black Holes
- The Explanation of Planck's Constant
- The Explanation of Increasing Velocity and Increasing Mass
- The Explanation of the Striking Parallel Between Newton's Law of Gravity and Coulomb's Law
- An Explanation of the Creation of the Universe
- The Explanation of Time and Time Dilation Effects
- The Grand Unification Theory
- Why Electrons Orbit Protons
- GOD! Has GOD been discovered!!!
- Universal Religion?
- The First ever Explanation of the Constant of Fine Structure!!!

Some Subjects found in Book 3:

THE EXPLANTION OF THE QUARK THEORY

- Assessment of the Quark Theory
- Creation of the 1/3 and 2/3 Charges
- Creation of the ±1 Charge, The ±2 Charge, and Spin
- Creation of the Up and Down Quarks
- Creation of the Strange and Charm Quarks
- Creation of the Top and bottom Quarks
- "The Four Layers of Matter"
- Neutrinos Explained
- How Particle Collisions Create New Particles
- "Tunneling"
- The Explanation of how Quarks Change "Flavor"
- The Explanation of the Law of the Conservation of Lepton Number
- Lepton Creation During the Decay of Positive and Negative Pions
- Neutrino Creation During the Decay of the positive Muon
- Neutrino Creation During the Decay of the Negative Muon
- The Collision Between a Proton and an Electron Anti-neutrino & a Proton and a Muon Anti-neutrino
- The Collision between a Neutron and an Electron Neutrino & the Neutron Muon Neutrino Collision
- The Decay of the Neutron and the Creation of the Anti-neutrino
- The Explanation of the Law of the Conservation of Baryons
- The Explanation of the Conservation of "Strangeness"
- Gauge Bosons are not Force Carriers Between Particles
- The Explanation of The Pauli Exclusion Principle
- The Explanation of the CPT Theorem
- The Motion of Photons and Particles through Electric and Magnetic Fields
- The Stability of Protons; The Instability of Mesons
- Eliminating Popular Misconceptions
- Major Problems with Today's Popular Theories

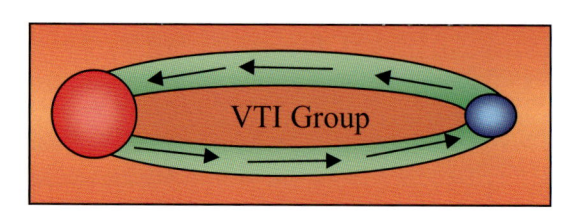

INDEX OF SCIENTIFIC DISCOVERIES FROM BOOK 3

- The Explanation of what Quarks are
- The Explanation of Quark Confinement
- The Explanation of the 1/3 & 2/3 Charges of Quarks
- The Explanation of ±2 Charge of Resonances
- The Explanation of the Up Quark
- The Explanation of the Down Quark
- The Explanation of the Strange Quark
- The Explanation of the Charm Quark
- The Explanation of the Bottom Quark
- The Explanation of the Top Quark
- The Explanation of the Muon
- The Explanation of the Tau
- The Explanation of the Electron, Muon, and Tau Neutrinos
- The Difference between Strong Force and Weak Force Creations
- The Explanation of how Quarks Change "FLAVOR"
- The Explanation of how Quarks Decay into other Types of Quarks
- The Explanation of "The Law of the Conservation of Lepton Number"
- The Explanation of the "Law" of the Conservation of Strangeness
- The Explanation of the Strange Quark's Extremely Long Lifetime
- The Explanation of the W Particle
- The Explanation of the Z Particle
- The Reason why Electric and Magnetic Fields exist in Photons and why They are at Right Angles to Each Other!
- The Explanation of the Pauli Exclusion Principle!
- The Explanation of the CPT Theorem
- The Explanation of the Asymmetric Parity of Neutrinos
- The Explanation of Gluons
- Dispelling the Myth of Gluons
- Dispelling the Myth of Gravitons
- Dispelling the Myth of the Higgs Bottom Particle
- The Explanation of the Three Color Charges in Quantum Chromodynamics
- The Explanation of the Law of The Conservation of Baryons
- The secret of quantum entanglement explained!